ORDER TODAY THE PROGRAM DISK

Investigate and develop understanding of modelling molecular structures with the Program Disk available for your IBM PC (and most compatibles).

In the disk, you will find:

–**BASIC** **programs** to do the following:
calculations for the H_2 problem
LCAO, VB and CI treatments
matrix diagonalization (for simple Hückel problems, etc...)
a PPP electron program
a rudimentary CNDO/2 all-valence electron program
– **End of Chapters problems with their solutions**

These are available in 3 different formats accessible by using:
• WordPerfect 5.1
• Postscript files which can be printed by a postscript printer
• PDF files which can be viewed and printed from the Adobe Acrobat Reader, available free via the Internet URL-http://www.adobe.com

Order the Program Disk today priced £9.99 including VAT and postage

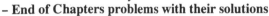

D1482176

Hinchliffe: Modelling Molecular Structures – Program Disk (ISBN 0 471 96491 3)

CALL IN YOUR ORDER ON:
+44 (0) 1243 843208

E-MAIL YOUR ORDER TO:
gbjwscr2@ibmmail.com

FAX IN YOUR ORDER ON:
+44 (0) 1243 842516

Please include:
1. ISBN, title and price
2. Full name, address and postcode

POST YOUR ORDER (USING THIS REPLY CARD)

Please send me........copies of **Hinchliffe: Modelling Molecular Structures** –Program Disk at £9.99 (including VAT and postage)

NAME ....

ADDRESS ....

POSTCODE ....

SIGNATURE ....

I enclose a cheque/P.O./Eurocheque for £ ....
(payable to John Wiley & Sons Ltd)

I wish to pay by credit card. Amount   £ ....

Diners Club   Barclaycard(Visa)   Americal Express   Mastercard
Card No ....   Expiry Date ....

Please keep me informed of new books in the following subject area(s)

....

Customer Service Department
John Wiley & Sons Limited
Distribution Centre
Shripney Road
Bognor Regis
Sussex
PO22 9SA
England

2/21/97

MODELLING MOLECULAR STRUCTURES

Wiley Tutorial Series in Theoretical Chemistry

Series Editors

D. Clary, University of Cambridge, Cambridge, UK
A. Hinchliffe, UMIST, Manchester, UK
D. S. Urch, Queen Mary and Westfield College, London, UK.

Previously Published Titles

Thermodynamics of Irreversible Processes: Applications to Diffusion
and Rheology

Gerard D. C. Kuiken

Published 1994, ISBN 0 471 94844 6

MODELLING MOLECULAR STRUCTURES

Alan Hinchliffe
UMIST, Manchester, M60 1D, UK

email Alan.Hinchliffe @umist.ac.uk

JOHN WILEY & SONS
Chichester · New York · Brisbane · Toronto · Singapore

Copyright © 1996 by John Wiley & Sons Ltd,
Baffins Lane, Chichester,
West Sussex PO19 1UD,
England

National   (01243 779777
International (+44) 1243 779777

Other Wiley Editorial Offices

John Wiley & Sons, Inc., 605 Third Avenue,
New York, NY 10158–0012, USA

Jacaranda Wiley Ltd, 33 Park Road, Milton,
Queensland 4064, Australia

John Wiley & Sons (Canada) Ltd, 22 Worcester Road,
Rexdale, Ontario M9W 1L1, Canada

John Wiley & Sons (SEA) Pte Ltd, 37 Jalan Pemimpin #05–04
Block B, Union Industrial Building, Singapore 2057

Library of Congress Cataloging-in-Publication Data

Hinchliffe, Alan
    Modelling molecular structures / Alan Hinchliffe.
        p.   cm. – (The Wiley tutorial series in theoretical
    chemistry)
        Includes bibliographical references and index.
        ISBN 0–471–95921–9 (hc : alk. paper). – ISBN 0–471–95923–5 (pbk.
    : alk. paper)
        1. Molecules–Models–Data processing.   I. Title.   II. Series.
    QD480.H56 1995
    541.2′2′0113536–dc20   95–17385
                                                                CIP

British Library Cataloguing in Publication Data

A catalogue record for this book is available from the British Library

ISBN 0 471 95921 9 (cloth)
ISBN 0 471 95923 5 (paper)

Typeset in 10/12pt Times by Alden Multimedia
Printed and bound in Great Britain by Bookcraft (Bath) Ltd, Midsomer–Norton, Avon
This book is printed on acid–free paper responsibly manufactured from sustainable forestation,
for which at least two trees are planted for each one used for paper production.

To my wife Joan

Series Preface

Theoretical chemistry is one of the most rapidly advancing and exciting fields in the natural sciences today. This series is designed to show how the results of theoretical chemistry permeate and enlighten the whole of chemistry together with the multifarious applications of chemistry in modern technology. This is a series designed for those who are engaged in practical research, in teaching and for those who wish to learn about the role of theory in chemistry today. It will provide the foundation for all subjects which have their roots in the field of theoretical chemistry.

How does the materials scientist interpret the properties of a novel doped-fullerene superconductor or a solid-state semiconductor? How do we model a peptide and understand how it docks? How does an astrophysicist explain the components of the interstellar medium? Where does the industrial chemist turn when he wants to understand the catalytic properties of a zeolite or a surface layer? What is the meaning of 'far-from-equilibrium' and what is its significance in chemistry and in natural systems? How can we design the reaction pathway leading to the synthesis of a pharmaceutical compound? How does our modelling of intermolecular forces and potential energy surfaces yield a powerful understanding of natural systems at the molecular and ionic level? All these questions will be answered within our series which covers the broad range of endeavour referred to as 'theoretical chemistry'.

The aim of the series is to present the latest fundamental material for research chemists, lecturers and students across the breadth of the subject, reaching into the various applications of theoretical techniques and modelling. The series concentrates on teaching the fundamentals of chemical structure, symmetry, bonding, reactivity, reaction mechanism, solid-state chemistry and applications in molecular modelling. It will emphasise the transfer of theoretical ideas and results to practical situations so as to demonstrate the role of theory in the solution of chemical problems in the laboratory and in industry.

D. Clary, A. Hinchliffe and D. S. Urch
June 1994

Preface

In the beginning, quantum chemists had pencils, paper, slide rules and log tables. It is amazing that so much could have been done by so few, with so little.

My little book 'Computational Quantum Chemistry' was published in 1988. In the Preface, I wrote the following:

> As a chemistry undergraduate in the 1960s . . . I learned quantum chemistry as a very 'theoretical' subject. In order to get to grips with the colour of carrots, I knew that I had to somehow understand
>
> $$\left| \int \Psi^*_k \sum_i \underline{r}_i \Psi_0 d\tau \right|^2$$
>
> but I really didn't know how to calculate the quantity, or have the slightest idea as to what the answer ought to be . . .

and I also drew attention to the new confidence of the late 1980s by quoting

> Today we live in a world where everything from the chairs we sit in to the cars we drive are firstly designed by computer simulation and then built. There is no reason why chemistry should not be part of such a world, and why it should not be seen to be part of such a world by chemistry undergraduates.

The book seemed to capture the spirit of the 1980s, and it became accepted as a teaching text in many universities throughout the world. In those days, computing was done on mainframes, scientific programs were written in FORTRAN, and the phrase 'Graphical User Interface' (GUI) was unknown.

Personal computing had already begun in the 1980s with those tiny boxes called (for example) Commodore PETs, Apples, Apricots, Acorns, Dragons and so on. Most of my friends ignored the fact that PET was an acronym, and took one home in the belief that it would somehow change their life for the better and also become a family friend. Very few of them could have written a 1024 word essay describing the uses of a home computer. They probably still can't.

What they got was an 'entry level' machine with a simple operating system and the manufacturer's own version of BASIC. There were no application packages to speak of, and there was no industry standard in software. Anyone who wrote software in those days would have nightmares about printers and disk files.

IBM (the big blue giant) slowly woke up to the world of personal computing, and gave us the following famous screen, in collaboration with MICROSOFT.

C:\ >

The DOS prompt.

Not very user-friendly!

Then came the games, and most older readers will recognise the Space Invaders screen shown below . . .

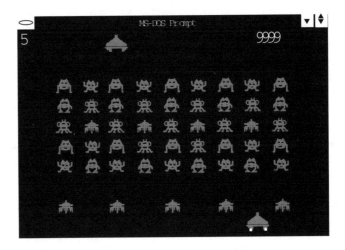

The Graphical User Interface was then born, courtesy of the the 'A' manufacturers such as Apple, Apricot, Amiga and Atari. Perhaps that is why so many of them went down the Games path. But they certainly left IBM behind.

These days we have Lemmings, Theme Park and SimCity. Many of them are modelling packages dressed up as games.

In the world of serious software, we soon saw the introduction of packages reflecting the three legs of the information technology trilogy

- word processing
- databases
- spreadsheets

and IT is now a well-established part of secondary education.

I don't want to bore you. As time went on, molecular modelling packages began to appear. Many ran under DOS (with the famous prompt screen above), but the more popular ran with GUIs on Apple Macs. Well, what happened is that MICROSOFT introduced WINDOWS, the famous graphical interface designed to protect users from DOS. There are now said to be more users of WINDOWS on IBM compatible PC's worldwide than all the other operating systems combined. But has all this actually changed our ability to understand molecules? Cynics will still argue that there have been no new major discoveries about molecular electronic structure theory since the heady days of the 1920's when Schrödinger, Pauli, Heisenberg and Dirac were active. Dirac said it all, in his oft-quoted statement

> Dirac's famous statement
>
> The underlying physical laws necessary for the mathematical theory of a large part of physics and the whole of chemistry are thus completely known, and the difficulty is only that exact application of these laws leads to equations much too complicated to be soluble

But computers and computing have moved along apace. This is especially true for personal computing, whereby powerful modelling packages are now available for everyday use. Most of these packages use molecular mechanics, and these come with brilliant graphics and excellent user-interfaces. Conformational problems involving protein strands that have been tackled using these packages are becoming common-place in the primary literature.

It is interesting to note that all the simple theories (such as Hückel π-electron theory) have now reappeared as options in these very same packages! Thus, very many scientists now routinely use computational quantum chemistry as a futuristic tool for modelling the properties of pharmaceutical molecules, dyestuffs and biopolymers. I wrote the original 'Computational Quantum Chemistry' Text as an Introduction for senior undergraduates and beginning postgraduates. True, the original edition had some flaws; reviewers pointed out that there was no need for a revision of the principles of quantum mechanics, however 'brief and breakneck' (and I quote).

It seemed to me that the time was ripe for a new text that would focus on recent applications, especially those reflected in current modelling packages for PCs. Hence this book!

I have also made available a disk containing the following teaching material

- BASIC programs to do the following:
  calculations for the H_2 problem; LCAO, VB and CI treatments
  Matrix diagonalization (for simple Hückel problems, etc.)
  a PPP π-electron program
  a rudimentary CNDO/2 all-valence electron program.
- End-of-Chapter problems.
- Solutions to all the End-of-Chapter problems.

In keeping with the spirit of the 1990s, I should mention that I have made use of the following PC packages in preparing this text.

ChemDraw Pro for Windows, by mass times velocity squared limited (mc^2), 46 Solent Road, London NM6 1TX (United Kingdom).

Chemintosh/ChemWindow, by Cherwell Scientific Publishing Limited, The Magdalen Centre, Oxford Science Park, OX4 4GA (United Kingdom).

MOBY Molecular Modelling on the PC, Version 1.5, by U. Höweler, Münster, FRG. Springer-Verlag GmbH & Co. KG Heidelberger Platz 3 D-14197 Berlin (Germany).

Mathcad 5.0 for Windows, by MathSoft, Inc. 101 Main Street, Cambridge, MA 02142.

DTMM3.0 for Windows, by M. James, C. Crabbe, John R. Appleyard and Catherine Rees Lay, Oxford University Press, Walton Street, Oxford OX2 6DP (United Kingdom).

I have also made use of the GAUSSIAN92 and GAUSSIAN94 packages. The literature citations are shown with the computer outputs.

Alan Hinchliffe
UMIST,
Manchester, 1995

Contents

0 Prerequisites **1**
    0.1   What on Earth is a Chapter 0? 1
    0.2   Basic Electrostatics 3
    0.3   Systems of Units. 5

1 Molecular Mechanics **9**
    1.1   Molecular Mechanics (The Concept) 12
    1.2   Molecular Mechanics (The Implementation). 17
    1.3   Protein docking 22

2 The Hydrogen Molecule-ion and Potential Energy Surfaces **25**
    2.1   The Born–Oppenheimer Approximation. 26
    2.2   The LCAO Model 28
    2.3   Integral Evaluation. 30
    2.4   Improving the Atomic Orbital 32
    2.5   The LCAO Approach 33

3 The Hydrogen Molecule **36**
    3.1   The Non-interacting Electron Model 38
    3.2   The Valence Bond Method 39
    3.3   Indistinguishability 40
    3.4   Electron Spin 41
    3.5   The Pauli Principle 42
    3.6   The H_2 Molecule 42
    3.7   Configuration Interaction 44
    3.8   The LCAO Molecular Orbital Method. 44
    3.9   Comparison of LCAO-MO and VB. 46
    3.10 Slater Determinants 46

4 The Electron Density **48**
    4.1   The General LCAO Case 51
    4.2   Population Analysis 52
    4.3   Density Functions and Matrices 54
    4.4   Density Functional Theory 55

5 Self Consistent Fields **56**
    5.1   The LCAO Procedure. 60
    5.2   Koopman's Theorem 64
    5.3   Open Shells 65
    5.4   Unrestricted Hartree Fock Theory. 66

6   Hückel Theory. **67**
  6.1   Examples. 69
  6.2   Extended Hückel Theory. 73
  6.3   'The Nightmare of the Inner Shells'. 75
  6.4   But What is the Hückel Hamiltonian? 77

7   Differential Overlap Models **78**
  7.1   The π-electron Models 79
  7.2   But Which χ_i are They? 85
  7.3   The 'All Valence Electron' Differential Overlap Models. 86
  7.4   Is There a Future for Semi-empirical Calculations? 95

8   Atomic Orbital Choice **97**
  8.1   Hydrogenic Orbitals 97
  8.2   Slater's Rules 98
  8.3   Clementi and Raimondi 99
  8.4   Gaussian Orbitals. 101
  8.5   Polarization and Diffuse Functions 109
  8.6   Literature Sources 110

9   An *Ab Initio* Package – GAUSSIAN92 . . . . **111**
  9.1   GAUSSIAN92. 112
  9.2   A GAUSSIAN92 Run 116
  9.3   The Hartree Fock Limit 124
  9.4   Open Shell Calculations 125
  9.5   A Guide to the Literature 133

10   Electron Correlation. **134**
  10.1   The Møller Plesset Method 137
  10.2   Configuration Interaction (CI). 144
  10.3   CID and CISD. 150
  10.4   Resource Consumption 151

11   The Xα Model. **152**
  11.1   The Atomic HF Problem 152
  11.2   Modelling Metallic Conductors 155
  11.3   Slater's Xα Method for Atoms 158
  11.4   Slater's Multiple Scattering Xα Method for Molecules. 159
  11.5   A Modern Implementation 160
  11.6   Conclusion. 166

12   Potential Energy Surfaces **167**
  12.1   A Diatomic Molecule 167
  12.2   Direct Differentiation: The Hellman Feynman Theorem 172
  12.3   Multiple Minima 173
  12.4   Potential Energy Surfaces that Depend on Several Variables 175
  12.5   Going Downhill on a Potential Energy Curve. 179

12.6   Analytical Differentiation Revisited   179
12.7   The Berny Optimization Algorithm   181
12.8   A Simple Example   181
12.9   Force Constants   186
12.10 Thermochemical Calculations   188
12.11 A Transition State   190

13   Primary Properties **192**
13.1   Electric Multipole Moments   193
13.2   Implication of Brillouin's Theorem   199
13.3   Quadrupole Moments   202
13.4   Higher Electric Moments   203
13.5   Electric Field Gradients   203
13.6   The Electrostatic Potential   206

14   Induced Properties **210**
14.1   Energy of a Charge Distribution in a Field   210
14.2   Induced Dipoles   212
14.3   Interaction Polarizabilities   222
14.4   The Hamiltonian   225
14.5   Magnetizabilities   226

15   Half a Dozen Applications **228**
15.1   Barriers to Internal Rotation and Inversion   228
15.2   Hydrogen-bonded Complexes   231
15.3   The Spherical Floating Gaussian (FSGO) Model   235
15.4   Isotropic Hyperfine Coupling Constants   237
15.5   Isodesmic Reactions   242
15.6   Crystalline Lattices such as Zeolites   245

References   248

Index   253

0 Prerequisites

0.1 WHAT ON EARTH IS A CHAPTER 0?

Let me tell you how things were in the heady days of the 1960s, when scientists (like me) and engineers first got our hands on computers. Computers were large beasts, and they consumed very many kilojoules (kilocalories in those days, or if you are a North American reader) per unit time. For this reason, they usually had a refrigeration plant, where the three resident engineers kept the milk for their coffee.

In those days, there were no packages such as GAUSSIAN92. All we had were rudimentary libraries containing matrix diagonalization 'subroutines' etc. The first step was to write your own code. We wrote programs in PIG (Programming Input General) or in an **autocode**, such as MERCURY AUTOCODE. By 'we', I mean the budding electrical engineers, the crystallographers and the very select group of quantum chemists who spent their nights and weekends running calculations on the Ferranti MERCURY. This machine occupied a large portion of the top floor of the Hicks Building in the University of Sheffield (UK).

What you would recognize as random access memory (RAM) was extremely limited on the Ferranti MERCURY, and so programs had to be segmented into 'CHAPTERS'. The first 'segment' of code was CHAPTER 0. CHAPTER 0 **had** to be executed before any other, and I hope that you will regard this chapter of my book in the same way. That is to say, you really must read this chapter before you begin the book in earnest.

The reason is quite simple. I am going to have to assume some prerequisite knowledge of the techniques of quantum chemistry. In particular, I think that you should know the following.

- The basic ideas of quantization (black body radiation, the photoelectron effect . . . that kind of thing).
- The basic postulates of quantum mechanics (stationary states, the Born interpretation of quantum mechanics which has $\psi^*\psi$ representing a probability and so on).

- Schrödinger's treatment of the hydrogen atom, and a familiarity with his famous equation for the stationary states

$$\frac{\partial^2 \psi}{\partial x^2} + \frac{\partial^2 \psi}{\partial y^2} + \frac{\partial^2 \psi}{\partial z^2} + \frac{8\pi^2 m}{h^2}(\epsilon - U)\psi = 0$$

  where ϵ is the electron energy and U the mutual potential energy of the electron and the nucleus.

- A knowledge of how to write the Schrödinger equation as an **eigenvalue equation**, involving the **Hamiltonian operator** H

$$H\psi = \epsilon\psi$$

- An appreciation that wavefunctions for molecules depend on the coordinates of **all** the particles present. I am going to write **capital** Ψ when dealing with many-particle systems, and lower case symbols such as ψ above when dealing with single particles such as the electron in a hydrogen atom or H_2^+ molecule-ion. I will also use χ to stand for atomic orbitals in a molecular treatment (e.g., the LCAO method).

- A knowledge of the **variation principle**. In particular, you should be familiar with variational integrals such as

$$\frac{\int \Psi^* H \Psi \, d\tau}{\int \Psi^* \Psi \, d\tau}$$

  I will normally treat all wavefunctions as real quantities, so you can forget about the complex conjugate * in equations of this kind.

- An appreciation that this variational integral has dimensions of energy, and that the variational energy is always greater or equal to the true energy, depending on whether the wavefunction is an approximation to the state of a given symmetry having the lowest energy, or exactly equal to it.

- A familiarity with the variation principle in its **linear** form, where we seek an approximate wavefunction as a linear combination

$$\Psi = c_1 \Psi_1 + c_2 \Psi_2 + \ldots + c_n \Psi_n$$

  and you should know that this leads to the so-called **secular equations**

$$\begin{pmatrix} H_{1,1} & H_{1,2} & \ldots & H_{1,n} \\ H_{2,1} & H_{2,2} & \ldots & H_{2,n} \\ \ldots\ldots \\ H_{n,1} & H_{n2} & \ldots & H_{n,n} \end{pmatrix} \begin{pmatrix} c_1 \\ c_2 \\ \ldots\ldots \\ c_n \end{pmatrix} = \epsilon \begin{pmatrix} S_{1,1} & S_{1,2} & \ldots & S_{1,n} \\ S_{2,1} & S_{2,2} & \ldots & S_{2,n} \\ \ldots\ldots \\ S_{n,1} & S_{n,2} & \ldots & S_{n,n} \end{pmatrix} \begin{pmatrix} c_1 \\ c_2 \\ \ldots\ldots \\ c_n \end{pmatrix}$$

- The concept of electron spin and the ideas of spin multiplicity.
- A familiarity with the idea of indistinguishability.
- A knowledge of the LCAO (linear combination of atomic orbitals) methodology.
- The **aufbau** principle, and the ability to write an electronic configuration for a first row diatomic such as O_2.

Also, you will need the concept of a **vector**. As you should know, vectors are quantities having magnitude and direction, and I am going to represent them by symbols such as $\underline{r}$ and $\underline{R}$. The point is that we need to describe the positions of electrons and nuclei in space, and the language of vectors turns out to be the most succinct.

0.2 BASIC ELECTROSTATICS

At various points throughout the text, I will need to refer to some of the basic concepts and results of classical electrostatics. This is a field of human endeavour that deals with the forces between electric charges at rest, the fields and electrostatic potentials produced by such charges, and the mutual electrostatic potential energy produced by a pair of charge distributions. To get us started, consider two point charges Q_A and Q_B, as shown in Figure 0.1.

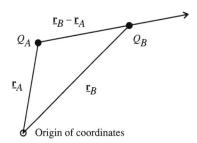

Figure 0.1   Construct needed to discuss the force between the point charges Q_A and Q_B

Point charge Q_A is located at vector position $\underline{r}_A$ and point charge Q_B is at vector position $\underline{r}_B$. The vector joining the positions of Q_A and Q_B is also shown; $\underline{r}_B - \underline{r}_A$ points in the direction **from A to B** as shown.

I am going to quote a number of results without proof, and I want you to bear them in mind:

- Point charge Q_A exerts a force $\underline{F}(Q_A$ on $Q_B)$ on Q_B given by

$$\underline{F}(Q_A \text{ on } Q_B) = \frac{Q_A Q_B}{4\pi\epsilon_0} \frac{(\underline{r}_B - \underline{r}_A)}{|\underline{r}_B - \underline{r}_A|^3}$$

- Point charge Q_A generates a field $\underline{E}$ at points in space, and in particular it generates a field at the position in space $\underline{r}_B$ where I originally placed Q_B. The field generated by Q_A exists irrespective of the presence or absence of Q_B, and I normally omit all mention of Q_B (Figure 0.2), and just focus attention on that particular point in space. The field at point $\underline{r}$ is written $\underline{E}(\underline{r})$

$$\underline{E}(\underline{r}) = \frac{Q_A}{4\pi\epsilon_0} \frac{(\underline{r} - \underline{r}_A)}{|\underline{r} - \underline{r}_A|^3}$$

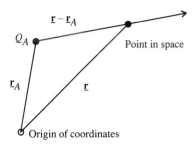

Figure 0.2   Construct needed to discuss the field generated by Q_A at a point in space

- Point charge Q_A generates an electrostatic potential at points $\underline{r}$ in space, given by

$$\varphi(\underline{r}) = \frac{Q_A}{4\pi\epsilon_0} \frac{1}{|\underline{r} - \underline{r}_A|}$$

The potential at $\underline{r}$ is written $\varphi(\underline{r})$.
- The **work done** in building up the charge distribution shown in the Figure 0.1, where the two point charges are initially at infinity, is called the **mutual electrostatic potential energy** $U(Q_A, Q_B)$ of Q_A and Q_B. For the pair of charges in my diagram it is

$$U(Q_A, Q_B) = \frac{Q_A Q_B}{4\pi\epsilon_0} \frac{1}{|\underline{r}_B - \underline{r}_A|}$$

When we deal with charge distributions rather than point charges, the definitions have to be generalized. What we do is to divide these charge distributions into differential charge elements $\rho(\underline{r})d\tau$, work out the contribution to the force, field, potential . . . made by this differential charge element, and then sum over them.

For example, if we were dealing with a charge distribution $\rho_A(\underline{r})$ instead of point charge Q_A (Figure 0.3) and we wanted to calculate the electrostatic field generated by ρ_A at the point $\underline{r}$ in space, then we would divide up the charge distribution ρ_A into differential charge elements $\rho_A d\tau$, and apply the basic formula for the electrostatic field. Here, $d\tau$ is a volume element. Finally, we would have to integrate over the coordinates of the charge distribution, in order to find the total field. This might not be easy!

$$\underline{E}(\underline{r}) = \int \frac{\rho_A(\underline{r}_A)}{4\pi\epsilon_0} \frac{(\underline{r} - \underline{r}_A)}{|\underline{r} - \underline{r}_A|^3} d\tau_A$$

Likewise to find the mutual potential energy of a pair of charge distributions $\rho_A(\underline{r}_A), \rho B(\underline{r}_B)$ we would have to calculate

$$U(\rho_A, \rho_B) = \int\int \frac{\rho_A(\underline{r}_A)\rho_B(\underline{r}_B)}{4\pi\epsilon_0} \frac{1}{|\underline{r}_B - \underline{r}_A|} d\tau_A d\tau_B$$

The integration would have to be over the volume of charge distribution A and charge distribution B.

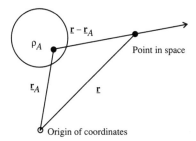

Figure 0.3   Construct needed to discuss the electric field generated by a charge distribution ρ_A

0.3 SYSTEMS OF UNITS

0.3.1   The Système International

It is usual these days to express all physical quantities in the system of units referred to as Système International–SI for short. SI units, which are recommended by the International Union of Pure and Applied Physics, are based on the metre, kilogram and the second as the fundamental units of length, mass and time. The SI electrical units are determined from these, via the fundamental constants ϵ_0 and μ_0, the permittivity and permeability of free space respectively. The ampere is defined in terms of the force between two straight parallel conductors placed a metre apart, and once this has been defined the coulomb must be such that one coulomb per second passes along a conductor if it is carrying a current of one ampere.

It turns out that the speed of light c_0, ϵ_0 and μ_0 are interrelated

$$c_0 = \frac{1}{\sqrt{\epsilon_0 \mu_0}}$$

and since 1983 the speed of light has been **defined** in terms of the distance that light travels in one second. It has the exact value

$$c_0 = 2.99792458 \times 10^8 \, \text{m s}^{-1}$$

We **are** going to be concerned with electrical and magnetic properties, so I had better put on record the fundamental force laws for stationary charges and steady currents. These are as follows.

$$\underline{F}(Q_A \text{ on } Q_B) = \frac{Q_A Q_B}{4\pi\epsilon_0} \frac{(\mathbf{r}_B - \mathbf{r}_A)}{|\mathbf{r}_B - \mathbf{r}_A|^3}$$

is the electrostatic force exerted **by** point charges Q_A **on** point charge Q_B, where $\underline{r}_A$ is the position vector of Q_A and $\underline{r}_B$ the position vector of Q_B. I discussed this above, and you should be aware that this force is exactly equal and opposite to the force exerted by Q_B on Q_A.

The corresponding force between two complete electrical circuits A and B is

$$\mathbf{F}(A \text{ on } B) = I_A I_B \frac{\mu_0}{4\pi} \oint \oint d\mathbf{l}_B \times (d\mathbf{l}_A \times \frac{\mathbf{r}_B - \mathbf{r}_A}{|\mathbf{r}_B - \mathbf{r}_A|^3})$$

which is much more complicated, because the integrations have to be done around the complete electrical circuits. The details do not matter, the point being this. Because ϵ_0 and μ_0 are interrelated, we are free to give one of them an arbitrary value, and in SI we choose arbitrarily to make

$$\mu_0 = 4\pi \times 10^{-7} \,\text{H m}^{-1}$$

0.3.2 Gaussian units

The most commonly used system apart from SI is the Gaussian system, sometimes called 'cgs' because it is based on the centimetre, gram and second. The unit of force is the dyne, and the unit of energy the erg.

There is usually no problem in converting between SI and Gaussian units, until we have to consider electrical phenomena. In cgs we take the proportionality constant in Coulomb's law to be unity (a number)

$$\mathbf{F}_{\text{cgs}}(Q_A \text{ on } Q_B) = Q_A Q_B \frac{(\mathbf{r}_B - \mathbf{r}_A)}{|\mathbf{r}_B - \mathbf{r}_A|^3}$$

and this means that derived equations have a different form. The unit of charge is called the **electrostatic unit** (esu). When two charges each of magnitude 1 esu are separated by a distance of 1 cm, each experiences a force of 1 dyne. The electric field is measured in statvolts cm^{-1}.

As a rule of thumb, be wary of equations which have an $(-e)^2$ but no $4\pi\epsilon_0$, and of equations that relate to highly symmetrical charge distributions but seem to have a 4π too many.

It gets worse with magnetic properties, and the Lorentz force

$$\mathbf{F} = Q(\mathbf{E} + \mathbf{v} \times \mathbf{r})$$

is written in such a way as to make the magnetic field $\mathbf{B}$ have the same dimensions as the electric field $\mathbf{E}$, namely force per unit length. In cgs units, the Lorentz force law becomes

$$\mathbf{F} = Q(\mathbf{E} + \frac{1}{c_0}\mathbf{v} \times \mathbf{r})$$

so as a final *aide-mémoire*, beware of magnetic equations that have a c_0 in them.

A quick conversion table is given in Table 0.1. It isn't comprehensive, but you should find it useful.

Table 0.1 Conversion factors between SI and cgs units
(taking the speed of light in free space as $3 \times 10^8 \, \mathrm{m\,s^{-1}}$)

Quantity	SI unit	cgs unit	No of Gaussian units in one SI unit
force	newton	dyne	10^5
energy	joule	erg	10^7
charge	coulomb	esu	3×10^8
current	ampere	$\mathrm{esu\,s^{-1}}$	3×10^8
potential	volt	statvolt	$1/300$
E	$\mathrm{volt\,metre^{-1}}$	$\mathrm{statvolt\,cm^{-1}}$	$1/(3\,10^4)$
B	tesla	gauss	10^4

Unfortunately, many texts dealing with molecular modelling still use cgs.

0.3.3 Hartree's system

One problem in molecular modelling is that large powers of 10 tend to appear in all the calculated quantities. Thus, for example, molecules have masses of typically 10^{-26} kg, ions have charges in multiples of 1.6×10^{-19} C and so on. In many engineering applications, it is normal to reduce an equation to dimensionless form, and this in essence is what I will now describe. It is **not mandatory**, but you will understand the output from many computer packages a lot more easily if you study this section!

The 'atomic unit of length' a_0 is equal to the radius of the first Bohr orbit for a hydrogen atom (and is usually called the bohr), whilst the 'atomic unit of energy', the hartree E_h, is twice the energy of a ground state hydrogen atom. This also works out as the mutual potential energy of a pair of electrons distance a_0 apart.

In terms of the electron rest mass and the electron charge, we find

$$a_0 = \frac{\epsilon_0 h^2}{\pi m e^2}$$

and

$$E_\mathrm{h} = \frac{e^2}{4\pi\epsilon_0 a_0}$$

Table 0.1 shows such 'atomic units'. The accepted values of the SI constants are themselves subject to minor experimental improvements, so we generally report the results of theoretical calculations as (e.g.) R = $50\,a_0$. It is expected that you will quote the value of the 'conversion factor' somewhere in your paper, usually as a footnote.

In many engineering applications, it is usual to rewrite the basic equations in dimensionless form. The same applies in quantum chemistry. For example, consider the electronic Schrödinger equation for a hydrogen atom

$$\left(-\frac{h^2}{8\pi^2 m}\nabla^2 - \frac{e^2}{4\pi\epsilon_0 r}\right)\psi_\mathrm{el}(\mathbf{r}) = \epsilon_\mathrm{el}\psi_\mathrm{el}(\mathbf{r})$$

(I regret any possible confusion caused by using the symbol ϵ for both energy and the permittivity of free space).

The energy and the distance r are both real physical quantities, with a measure and a unit. If we define the dimensionless variable $r_{red} = r/a_0$ as the ratio of the length to a_0, then r_{red} is a dimensionless variable. The idea is to rewrite the electronic equation in terms of the reduced variables; to do this, we substitute $r = r_{red} a_0$, $\epsilon_{el} = \epsilon_{red} E_h$, etc. and what we find is a much simpler, dimensionless equation

$$\left(-\frac{1}{2}\nabla^2_{red} - \frac{1}{r_{red}}\right)\psi_{el,red}(\mathbf{r}_{red}) = \epsilon_{el,red}\,\psi_{el,red}(\mathbf{r}_{red})$$

I am afraid that it is common practice for people to forget about the fact that such variables are dimensionless, write down equations such as the electronic Schrödinger equation above, and quote the results as energies, etc.

Even worse is the confusion regarding the wavefunction itself. The Born interpretation of quantum mechanics tells us that $\psi^*(\mathbf{r})\psi(\mathbf{r})d\tau$ represents the probability of finding the particle with coordinates $\mathbf{r}$, described by wavefunction $\psi(\mathbf{r})$, in volume element $d\tau$. Probabilities are real numbers, and so the dimensions of $\psi(\mathbf{r})$ must be $(\text{length})^{-3/2}$. In the atomic system of units, we take the unit to be $a_0^{-3/2}$.

For your guidance, Table 0.2 will help you convert into SI the results of theoretical calculations. The first column gives the physical quantity. The second column shows the usual symbol. The third column gives X, the collection of physical constants corresponding to each quantity. This collection is not unique, but the value given in the fourth column **is** unique.

Table 0.2

Physical quantity	Symbol	X	Value of X
length	l	a_0	$5.2918 \times 10^{-11}\,\text{m}$
mass	m	m_e	$9.1095 \times 10^{-31}\,\text{kg}$
energy	E (or ϵ)	E_h	$4.3598 \times 10^{-18}\,\text{J}$
charge	Q	e	$1.6022 \times 10^{-19}\,\text{C}$
electric dipole moment	p_e	ea_0	$8.4784 \times 10^{-30}\,\text{C m}$
electric quadrupole moment	Θ	ea_0^2	$4.4866 \times 10^{-40}\,\text{C m}^2$
electric field	E	$E_h e^{-1} a_0^{-1}$	$5.1423 \times 10^{11}\,\text{V m}^{-1}$
electric field gradient	$-V_{zz}$	$E_h e^{-1} a_0^{-2}$	$9.7174 \times 10^{21}\,\text{V m}^{-2}$
magnetic induction	B	$(h/2\pi)e^{-1}a_0^{-2}$	$2.3505 \times 10^5\,\text{T}$
electric dipole polarizability	α	$e^2 a_0^2 E_h^{-1}$	$1.6488 \times 10^{-41}\,\text{C}^2\,\text{m}^2\,\text{J}^{-1}$
magnetizability	ξ	$e^2 a_0^2 m_e^{-1}$	$7.8910 \times 10^{-29}\,\text{J T}^{-2}$

1   Molecular Mechanics

Congratulations, and welcome to the text. You are about to begin your study of 'Modelling'. At the outset, I must admit that this book is not quite as comprehensive as the title might suggest. You won't meet every type of molecule, every model that has been tried over the years nor every molecular structure.

To get us going, we need to establish a good foundation in the mathematical treatment of those scientific models which can be used to study molecules and molecular structures. The word 'model' has a special meaning in science. It does **not** mean sitting down immediately at a PC and drawing on the screen, although modellers may spend some of their time on that activity. It means having a set of mathematical equations which are capable of representing accurately the chemical phenomenon under study. Thus, we can have a model of the UK economy just as we can have a model of a Vauxhall Astra or of a naphthalene molecule.

Why do we want to model molecules? Chemists are interested in the behaviour of electrons around the nuclei and, of course, about how the electrons rearrange in a chemical reaction; this is what chemistry is all about. Thompson realized this in 1897, when he tried to develop an electronic theory of valence. He was quickly followed by Lewis, Langmuir and Kössel, but their models all suffered from the same defect; they tried to treat the electrons as if they were point electric charges at rest. The year 1926 was an exciting one. Schrödinger, Heisenberg and Dirac, all working independently, solved the hydrogen atom problem. Schrödinger's method, which we now call **wave mechanics**, is the version that you are all familiar with. The only cloud on the horizon was summarized by Dirac in his famous statement, as quoted in the preface.

I think that Dirac (1929) was warning us that solution of the equations of quantum mechanics for everyday problems was going to be horrendous. He was, of course, right! The kind of problems that people could tackle successfully in the early days were very simple and qualitative. For example:

- why is the H atom stable, and what are its allowed energy levels?
- why is the hydrogen molecule ion H_2^+ stable, and what should its bond length be?
- why is methane tetrahedral?
- why is the bond angle in water 108°?

We are going to have a rerun of some of these in later chapters.
Fifty years on from there, Clementi (1973) saw things differently

> We can calculate everything

Frank Boys (1950) saw things in a little more perspective when he said

> It has thus been established that the only difficulty which exists in the evaluation of the energy and wavefunction of any molecule . . . is the amount of computing necessary

These days, even the simplest problems discussed in the market leader literature (such as *The Journal of The American Chemical Society*, the *Journal of Molecular Structure (THEOCHEM)*, *Chemical Physics Letters*, *Electronic Journal of Quantum Chemistry*, and *Journal of Chemical Physics*) are much more sophisticated.

- What are the bond lengths and bond angles in nicotine (Figure 1.1)? And can we visualize its reactivity in simple terms?

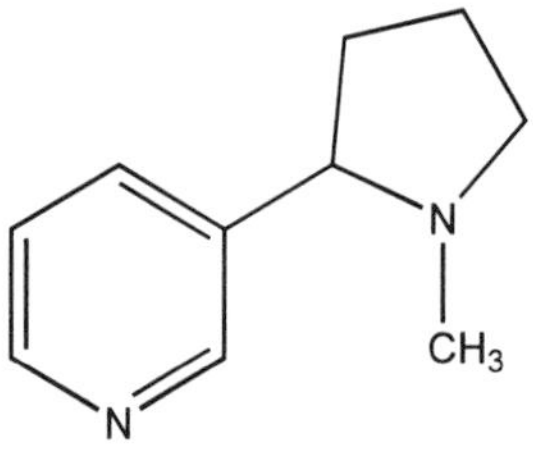

Figure 1.1   Nicotine

- What is the three-dimensional, geometrical conformation of isoflupredone (a molecule widely used in vet science) (Figure 1.2)?

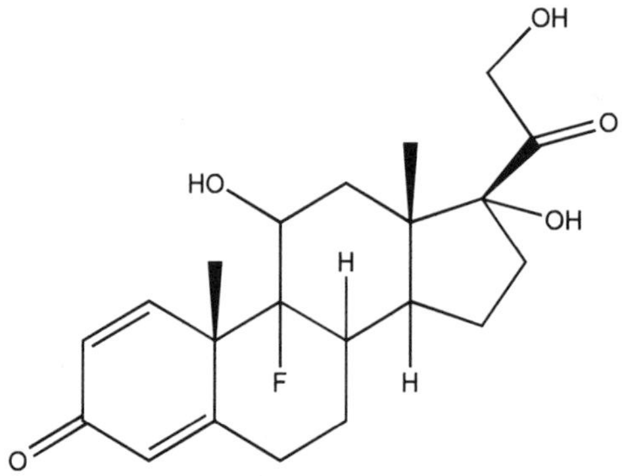

Figure 1.2   Isoflupredone

- Naphthalene (Figure 1.3) has a high symmetry, and its electric dipole moment is zero. When it is attacked by a charged reagent such as NO_2^+, the charge density rearranges and an electric dipole can be induced. A measure of this rearrangement is offered by the electric dipole polarizability, which relates the induced dipole to the external field generated by the attacking reagent. Polarizabilities are very hard to measure experimentally, so can we calculate them instead? And if so, how accurate are our predictions likely to be?

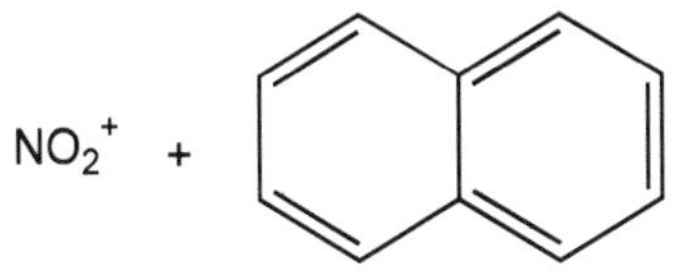

NO_2^+ +

Figure 1.3   Nitration of naphthalene

- What is the reaction path, transition state and activation energy for a typical chemical reaction such as shown in Figure 1.4.

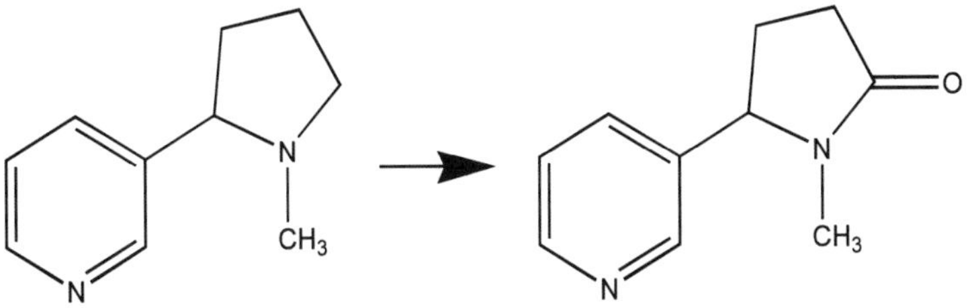

Figure 1.4   A simple organic chemical reaction

All these phenomena (and more) have something in common. They are to do with particles in motion, in this case electrons, neutrons and protons. But they could equally well have been colliding billiard balls and express trains. Or could they? Perhaps I should remind you that **mechanics** is the branch of mathematics dealing with movement. There are three branches of mechanics:

- **classical mechanics** normally deals with things in the everyday world: accelerating sports cars, bodies sliding down inclined planes . . . that kind of thing;
- **relativistic mechanics** normally deals with situations where one particle is moving at a speed of the order of magnitude of that of light, with respect to another particle;
- **quantum mechanics** normally deals with situations where small particles are involved.

You were probably taught early in your professional career that skills in quantum chemistry are a prerequisite for a study of atomic and molecular phenomena. I must tell you that this isn't completely true. Some molecular phenomena can be modelled very accurately indeed using **classical** mechanics, but for many other phenomena we do indeed have to take account of the intrinsic quantum behaviour of nature.

1.1   MOLECULAR MECHANICS (THE CONCEPT)

To get us started, let's consider a very simple **classical** model for vibrating molecules, and one that you will certainly have met before. A vibrating molecule can be modelled successfully if it is imagined to be made from balls on the ends of springs (Figure 1.5). The motion is then treated classically. The first mention of this technique in the literature seems to be given by Andrews (1930). The first practical calculation seems to be that of Westheimer and Meyer (1946). It turns out that the model is a good one, and there is a direct payoff. This simple model is used widely to predict the conformations of large biological and pharmaceutical molecules. In fact, the phrase 'Molecular Modelling' tends to mean just this to many people.

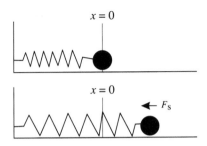

Figure 1.5   A particle of mass **m** joined to a spring. F_s is the restoring force

1.1.1   The simple harmonic oscillator

The simplest possible case that we can study for molecular vibrational motion is that of a particle with mass m attached to a spring. The particle is set in motion and it vibrates about some fixed position. Call the spring extension x. Experience shows that, when the spring is extended, the particle experiences a restoring force which I have shown as F_s in Figure 1.5. The subscript 's' means spring. Likewise, if the spring is compressed, there is a restoring force which acts to move the particle back to its position of rest where $x = 0$. For very many situations, it turns out that the force

$$F_s = -k_s x$$

depends on the extension and this is known as **Hooke's Law**. The proportionality constant k_s is called the **force constant**.

    Not all materials obey Hooke's Law, and even those materials which do, show deviations for a large extension. But it is a good place to start our study of molecular modelling, as it turns out that molecules vibrate in much the same way as particles attached to springs.

You should know that

$$\text{force} = \text{mass} \times \text{acceleration (one of Newton's laws)}$$

and so the motion of the particle is described by Newton's equation

$$m\frac{\mathrm{d}^2 x}{\mathrm{d}t^2} = -k_s x$$

This is a second-order differential equation with the standard solution

$$x = A\sin(\sqrt{\frac{k_s}{m}}\,t) + B\cos(\sqrt{\frac{k_s}{m}}\,t)$$

A and B are constants of integration; we have to know the behaviour of the system at two different times in order to find them. The constants A and B need not concern us here. The quantity $(k_s/m)^{\frac{1}{2}}$ occurs again and again in the treatment of vibrational motion; it has units of radians $(\text{time})^{-1}$ and so is an angular frequency. We call

$$\upsilon_{cl} = \frac{1}{2\pi}\sqrt{\frac{k_s}{m}}$$

the **classical vibration frequency** of the system.

1.1.2 The potential energy U

You are probably used to the idea of conservation of energy; the kinetic and the potential energy U add to give a constant which I will call ϵ. How do we find the potential energy U, given the force? That is, we want to be able to find a U such that the total energy is constant

$$\frac{1}{2}mv^2 + U = \text{constant } (\epsilon)$$

but how do we relate U to F_s? To discover the link, we differentiate each side of the total energy equation with respect to time

$$\frac{\mathrm{d}}{\mathrm{d}t}(\frac{1}{2}mv^2 + U) = 0$$

$$mv\frac{\mathrm{d}v}{\mathrm{d}t} + \frac{\mathrm{d}U}{\mathrm{d}t} = 0$$

$$m\frac{\mathrm{d}x}{\mathrm{d}t}\frac{\mathrm{d}v}{\mathrm{d}t} + \frac{\mathrm{d}U}{\mathrm{d}x}\frac{\mathrm{d}x}{\mathrm{d}t} = 0$$

$$\frac{\mathrm{d}x}{\mathrm{d}t}(m\frac{\mathrm{d}v}{\mathrm{d}t} + \frac{\mathrm{d}U}{\mathrm{d}x}) = 0$$

$$\frac{\mathrm{d}x}{\mathrm{d}t}(F + \frac{\mathrm{d}U}{\mathrm{d}x}) = 0$$

where I have used Newton's law in the final step. This shows that, for a one-dimensional problem

$$F = -\frac{dU}{dx}$$

and that is how we find a force from a potential; to find a potential from a force we have to integrate

$$U = -\int F dx$$

In the case of a Hooke's law spring where $F_s = \frac{1}{2}k_s x$, you should be able to prove that $U = \frac{1}{2}k_s x^2$ + a constant of integration, which I will write as U_0.

> **Hooke's Law potential**
> $$U = U_0 + \frac{1}{2}k_s x^2$$

We can usually take $U = 0$ when $x = 0$, which gives $U_0 = 0$.

In the case of a three dimensional problem where $\underline{\mathbf{F}}$ is a vector that depends on three dimensions, the relationship between force and potential is a little more complicated. I will discuss this later in the chapter.

Is the definition of the Hooke's law potential I have just given you consistent with the idea that energy is conserved? Well, you should be able to differentiate the expressions I have given you for x and substitute x and dx/dt into the energy expression. You will find that the total energy really is constant.

When dealing with molecular vibrational problems where the potential obeys Hooke's law, we talk about the **harmonic approximation**. But how can a molecule be a spring? What **is** the spring and how can we find the force constant? As we will see later, the nuclei vibrate under the influence of the electrons. And in order to come to grips with the force constant, I need to remind you of some basic vibrational spectroscopy.

1.1.3 An aside . . . the quantum mechanical treatment

You will probably be aware of the full quantum mechanical treatment of diatomic vibrational motion? What happens if we try and solve the relevant Schrödinger equation for one-dimensional vibrational motion

$$\frac{d^2\psi}{dx^2} + \frac{8\pi^2 m}{h^2}(\epsilon - U)\psi = 0$$

where ψ is a vibrational wavefunction and U the vibrational potential energy ($\frac{1}{2}k_s x^2$ for the one-dimensional Hooke's law problem). Solution of the vibrational Schrödinger equation is not easy, but it **can** be done given a knowledge of things called 'special functions of mathematical physics'. We find different results from the classical ones in that

- the vibrational energy is quantized with
- quantum numbers $v = 0, 1, 2, \ldots$
- quantum mechanical energy levels are given by

$$\epsilon_{qm} = \frac{h}{2\pi}\left(v + \frac{1}{2}\right)\sqrt{\frac{k_s}{m}}$$

Spectroscopists usually talk in terms of 'wavenumbers', which are energies/hc_o, and they refer to the classical vibrational wavenumber as ω_e, given by

$$\omega_e = \frac{1}{2\pi c_o}\sqrt{\frac{k_s}{m}}$$

Analysis of a vibrational spectrum gives the force constant, k_s. But real diatomic molecules do not vibrate as if they were simple particles on the ends of classical springs, and professional spectroscopists would scoff at the idea of Hooke's law. They would be more concerned with matching as exactly as possible the experimental vibrational energy levels for a diatomic molecule, and instead of the Hooke's law potential they might write a Morse potential

$$U = D_e(1 - \exp(-ax))^2$$

where D_e and a are two parameters which give a more reasonable description of the potential [D_e is the depth of the potential energy well and $a = \frac{1}{2}\omega_e(2m/D_e)^{1/2}$]. Alternatively, they might try a potential energy function that was much more sophisticated.

But surely we can use quantum mechanics to calculate diatomic potential energy curves these days? Well, yes we can, and to look ahead a chapter the Figure 1.6 shows the result of a simple quantum chemical calculation on H_2^+ at the LCAO level of theory.

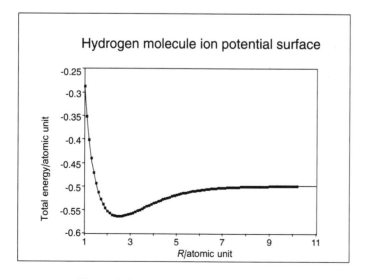

Figure 1.6 LCAO calculation on H_2^+

The potential energy curve does not seem at first sight to bear much resemblance to a Hooke's law curve. As the nuclei get close together, the dominant feature is repulsion between the nuclei and so the curve rises steeply. For large internuclear separations, the potential energy becomes that of the component atoms at infinity (in this case, H + H^+). For small vibrations about the equilibrium position though, the simple harmonic approximation is a good one. You can see this from Figure 1.7, where I have fitted points in the vicinity of the equilibrium bond length to a Hooke's law potential. For more accurate work, or for a study of the vibrations of large amplitude about the equilibrium distance, the simple harmonic approximation is a good place to start, but incomplete.

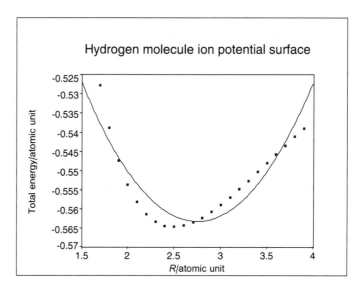

Figure 1.7 H_2^+ LCAO calculation, harmonic fit to potential energy curve near the minimum

1.1.4   Three dimensional models

One-dimensional problems are fine, but molecules are three dimensional and so we need to cater for the case where the force is three dimensional. How do we relate the force to the potential? You should know how to represent forces as vectors. I am going to write such vector quantities as $\underline{\mathbf{A}}$, and you will have come across the idea of unit vectors and the resolution of a vector into components along cartesian axes before.

In three dimensions, we write a Hooke's Law restoring force as

$$\underline{\mathbf{F}}_s = F_x \underline{\mathbf{e}}_x + F_y \underline{\mathbf{e}}_y + F_z \underline{\mathbf{e}}_z$$

How do we find an expression for the potential energy U in this case?

I will leave you to repeat the derivation given above for the one-dimensional case, but you must remember that the force and the potential energy are now both functions of the three variables, x, y and z. Where you had a dF/dx in the derivation, you must put a partial derivative $\partial F/\partial x$. You will find that

$$\mathbf{F}_s = -\left(\frac{\partial U}{\partial x}\mathbf{e}_x + \frac{\partial U}{\partial y}\mathbf{e}_y + \frac{\partial U}{\partial z}\mathbf{e}_z\right)$$

and this is often written as $\mathbf{F}_s = -\text{grad}\,U$ or $\mathbf{F}_s = -\nabla U$ where ∇ is the vector operator

$$\nabla = \left(\frac{\partial}{\partial x}\mathbf{e}_x + \frac{\partial}{\partial y}\mathbf{e}_y + \frac{\partial}{\partial z}\mathbf{e}_z\right)$$

The force is given by the negative of the gradient of the potential.

1.2 MOLECULAR MECHANICS (THE IMPLEMENTATION)

The idea that a molecule can be modelled as a collection of balls joined together with springs is as old as the hills. The molecular mechanics model capitalizes on this idea by seeking to express the intramolecular potential energy as a sum of terms that comprise the differences between actual and reference values of the molecular geometries.

Let's see how it might be done. Consider a diatomic molecule such as $^{12}C^{16}O$ in its electronic ground state, where spectroscopic studies have given the fundamental vibrational wavenumber ω_e as $2169.8\,\text{cm}^{-1}$ and the equilibrium bond length as $112.8\,\text{pm}$ (from Herzberg and Huber, 1979). Assume for the moment that the simple harmonic model is adequate, and that the force constant for such a simple harmonic oscillator is therefore $4\pi^2 c^2 \omega_e^2 \mu$ (where μ is the reduced mass, $1.139 \times 10^{-27}\,\text{kg}$), which gives $k_s = 1901.5\,\text{N}\,\text{m}^{-1}$. The potential energy for the $^{12}C^{16}O$ vibrational motion is therefore

$$U(CO) = \tfrac{1}{2}1901.5\,\text{N}\,\text{m}^{-1}\,(R_{CO} - 112.8\,\text{pm})^2$$

and U has a minimum at $R_e = 112.8\,\text{pm}$.

Minimum of U gives the equilibrium geometry

You have probably also noticed that

Second derivative of U gives the force constant

Likewise, a calculation on $^{12}C^{32}S$ gives

$$U(CS) = \frac{1}{2}849.0\,\text{N}\,\text{m}^{-1}\,(R_{CS} - 153.5\,\text{pm})^2$$

and so we might try a sum of these two, in order to describe the vibrations of the linear $^{16}O^{12}C^{32}S$ molecule. That is,

$$U(OCS) = U(OC) + U(CS)$$

or in explicit terms

$$U(\text{OCS}) = \tfrac{1}{2} 1901.8 \, \text{N m}^{-1} \, (R_{\text{CO}} - 112.8 \, \text{pm})^2 + \tfrac{1}{2} 849.0 \, \text{N m}^{-1} \, (R_{\text{CS}} - 153.5 \, \text{pm})^2$$

where the total potential energy $U(\text{OCS})$ now depends on the two independent variables R_{CH} and R_{CS}.

The problem of how to find those values of R_{CH} and R_{CS} that make U a minimum in this case is very easy; the minimum is still at $R_{\text{OC}} = 112.8 \, \text{pm}$ and $R_{\text{CS}} = 153.5 \, \text{pm}$, because the molecular potential is a sum of squares (with each term representing an independent bond). That is to say, the vibrations are completely independent of each other – they aren't **coupled** (to use the correct technical language). This isn't very exciting chemically, as we would like to use the model to **predict** the bond lengths, and we would anticipate that they would be different in the triatomic molecule. We refer to models where we write the total potential energy in terms of bond lengths, bond angles and so on as **valence force field** models.

Remember that we have constrained the OCS molecule to be linear, but something else is obviously missing!

To a spectroscopist, the problem is now one of how a force field should be determined from a set of experimental vibrational frequencies. This problem is well understood, and is discussed in the definitive text books (such as that by Wilson, Decius and Cross). The main difficulty is this; even for our simple, linear molecule, and even after assuming the harmonic approximation, the potential function will have several **cross terms** that I missed from the simple expression above. The true $U(\text{OCS})$ expression given above will actually be

$$\begin{aligned}
U(\text{OCS}) &= \tfrac{1}{2} k_{\text{OC}} (R_{\text{OC}} - R_{\text{e,OC}})^2 + \tfrac{1}{2} k_{\text{CS}} (R_{\text{CS}} - R_{\text{e,CS}})^2 \\
&\quad + \tfrac{1}{2} k_{\text{OC,CS}} (R_{\text{OC}} - R_{\text{e,OC}})(R_{\text{CS}} - R_{\text{e,CS}}) + \dots
\end{aligned} \tag{1}$$

where k $_{\text{OC,CS}}$ is called an **off-diagonal** force constant. It couples together the OC and the CS vibrations.

In fact, for a molecule of n atoms there are $(3n - 6)(3n - 5)$ independent force constants, which far exceeds the number of vibrational frequencies. A complete determination of all the force constants requires the analysis of the spectra of many isotopically substituted molecules. Many of these off-diagonal terms are very small, and spectroscopists have developed systematic simplifications to the force field in order to make as many of the off diagonal terms as possible vanish.

The key study for the development of molecular mechanics was the study by Schnachtschneider and Snyder, who showed that transferable force constants can be obtained when a few off-diagonal terms are **not** neglected. These authors found that cross-terms are usually largest when neighbouring atoms are involved.

The **Urey-Bradley** force field is similar to the valence force field except that non-bonded interactions are taken into account.

In molecular mechanics, the aim is quite different; it is to arrive at a set of transferable force constants k_s and reference equilibrium geometries R_e, that are good for making predictions on related molecules.

Two things should now be clear, even for such a simple case as linear $^{16}\text{O}^{12}\text{C}^{32}\text{S}$;

- There is a need to describe the non-bonded interaction(s).
- The C atom in OCS is obviously chemically different from the C atom in ethane and from the C atom in ethyne, and it is necessary to take careful account of the chemical environment of each atom. Is a C atom sp^2 hybridized or sp^3?

The latter point is very important, as it implies that a given atom will have a different set of reference bond lengths and force constants depending on its chemical environment. The idea then is to treat these force constants, the reference equilibrium distances and virtually everything else as 'parameters', which have to be fixed by reference to some molecular properties.

You can imagine the difficulty of trying to set up a system of reliable and transferable parameters for a large molecule containing many different types of atoms, and it should come as no surprise to find that the original Molecular Mechanics calculations were performed on hydrocarbons. The aim is invariably to predict the molecular geometry by minimizing the intramolecular potential energy, and the original set(s) of parameters were gradually extended over the years to cover the cases of substituted hydrocarbons.

So, consider a simple 'real' molecule such as aspirin. There is a conjugated system which is presumably planar, but the side chains obviously interact with each other. We have to allow at least for the following contributions to the intramolecular potential (U):

- a force constant k_s for each bond stretching motion;
- a force constant k_θ for each bond bending motion, where θ is the bond angle;
- a force constant k_ϕ for each dihedral angle, where ϕ is the dihedral angle. This term depends on the local symmetry; if we were dealing with CH_3 group rotations as in ethane, we would probably be tempted to put $U(CH_3 \text{ rotation}) = (k_0/2)[1 + \cos(3\omega)]$ where ω is the angle between CH groups on different centres. For systems of lower symmetry we would seek a potential which depended on the local axis of symmetry, and if this were of the order n then we would write $U = (k_0/2)S_n \cos(n\omega)$, where S_n defines the barrier to rotation.

The total potential energy U would be (so far)

$$U = \frac{1}{2}\sum k_{s,i}(R_i - R_{e,i})^2 + \frac{1}{2}\sum k_{\theta,i}(\theta_i - \theta_{e,i})^2 + \frac{1}{2}\sum k_{\omega,i}S_n \cos(n\omega_i)$$

where I have assumed that the stretches and bends can be well described by a quadratic term. R_i is the length of bond i (and is unknown at the start of the calculation) and $R_{e,i}$ the reference equilibrium bond distance parameter for that bond.

I mentioned earlier that whilst spectroscopist's force fields have to describe accurately a given molecule's vibrations, molecular mechanics (MM) force fields have to be transferable from molecule to molecule, and it was found that extra terms are needed in the potential energy sum above, in order to make this happen:

- non-bonded interactions, which are written as Lennard–Jones 12–6 contributions;
- and finally, some authors also include ionic contributions, in order to account for polar groups. You will remember from Chapter 0 that the mutual potential energy of

a pair of electric charges Q_A and Q_B distance R_{AB} apart is given by

$$U = \frac{Q_A Q_A}{4\pi\varepsilon_0 R_{AB}}$$

In order to make and use a MM model for a large molecule, we need at least three things:

- a reliable force field;
- a trial molecular geometry, to get us started;
- a decent algorithm for finding the minimum of a function of many (which can be **hundreds**) variables.

I haven't the space to discuss the various force fields in detail; the most famous empirical force field is probably MM2 (Molecular Mechanics 2) by Allinger and coworkers, and I refer you to the definitive text by Burkert and Allinger (1982) for more details. An improved version, MM3, has been published by Allinger, Yuh and Lii (1989), and the AMBER (Weiner and Kollman 1981) and CHARMm programs (Brooks et al 1983) were developed to deal with proteins and nucleic acids. A feature of the AMBER force field is that it treats CH groups as 'composite' atoms.

Many of the available packages have libraries of starting molecules and molecular fragments. In the early days of molecular modelling, people had lots of fun taking X-ray pictures of mechanical models or projecting shadows of these ball-and-stick models onto screens. At the end of the day, it is necessary to have the (x,y,z) coordinates of each atom in the molecule. There are several sources of X-ray data to get you started, of which the most famous two seem to be,

The Cambridge Structural Database System
Cambridge Crystallographic Data Centre
University Chemical Laboratory
Lensfield Road
CAMBRIDGE CB2 1EW, UK

Brookhaven Protein Data Bank
Brookhaven National Laboratory
c/o Ms F. C. Bernstein
Chemistry Department
Upton
New York
NY 11973, USA

The MOBY 1.5 modelling package comes complete with a selection of protein structures from the Brookhaven source. Alternatively, the starting geometry can usually be generated from a 'standard' geometry of averaged experimental data. The CONCORD program (Pearlman 1987) uses artificial intelligence to generate three dimensional geometrical data from such standard data and details of the atomic connectivities.

I am going to say a great deal in a later chapter about the methods for geometry searching, but I would just like to note here that the prediction of a reliable geometry for a small protein is asking rather a lot! There are a very large number of possible conformers.

There are a number of MM packages around, and several have been marketed for PC users. Amongst the packages on my PC, I have Desktop Molecular Modeller Version 3 (DTMM3). This package has been around for a long time as a DOS application, but the version I have is the very latest Microsoft Windows one. As with many of these packages, there is a 'molecules library'; if your target molecule isn't included, then it is possible to construct a starting geometry from the contents of the library and a few fragments. This is much simpler than starting from scratch (i.e. from earth, air, fire and water).

To give a very simple example, and show just how easy these packages are in operation, I decided to try aspirin which is a very simple pharmaceutical molecule. It is simple from a modelling point of view because of the benzene ring, which keeps the molecule largely planar. (As I will explain in a later chapter, benzene rings do cause problems in molecular modelling calculations). What I did was to load **benzene** from the library. I then had to delete a couple of adjacent hydrogens and begin building up my target molecule. You add an O atom and then a C atom in order to make up the $OCOCH_3$ group, and then you add a C and an O and a CH_3 to make up the $COCH_3$ group.

The Windows screen then looks like Figure 1.8

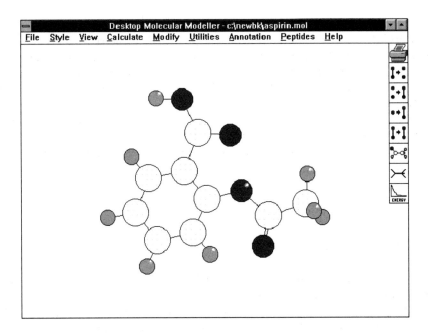

Figure 1.8   Initial Windows screen from DTMM3 for aspirin

The next step is the geometry optimization, which I will cover in detail in later chapters. As with all Windows applications, we click on the correct icon to choose that option, and after a while we get the structure shown in Figure 1.9.

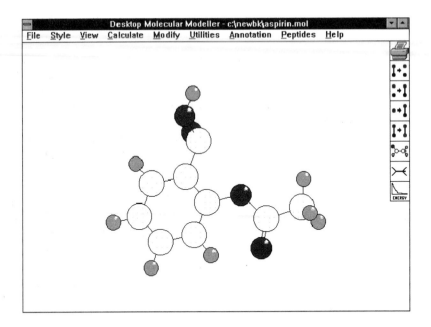

Figure 1.9   DTMM3 Windows screen for aspirin, after geometry optimization

The problem now is to decide whether the predicted geometry is the 'true' minimum. It might just be that there are many other structures having lower energy than this one. If the benzene ring had been a cyclohexane ring, you would have probably expected to find very many conformers with slightly different energies. That is a story for another chapter.

Apart from the predicted geometry, these MM packages will often calculate structural features such as the surface area or the molecular volume. Quantities such as these are often used to investigate relationships between molecular structure and pharmacological activity.

1.3   PROTEIN DOCKING

One great advantage of the Molecular Mechanics model is that it can be applied to large molecules on your average PC.

Apart from single molecule structure calculations, many researchers in the life sciences are concerned with the interaction between two (or more) large molecules. For example, the interaction between two amino acids, and we normally talk about **Protein Docking** when we study this phenomenon.

Let me give you an example, on which to end this Chapter. The screen grab below shows lysine and glutamic acid. I just loaded them in turn from DTMM3's library. There is nothing to be read into their relative orientations and positions in space.

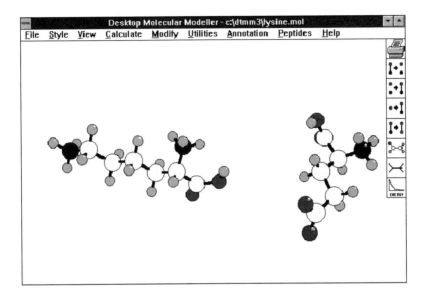

Figure 1.10   Protein docking

It is widely believed that such amino acids can help stabilize protein structures, because of the strong electrostatic interactions between their side chains. The idea is to investigate how these two amino acids interact with each other as they approach their equilibrium geometry, and we normally take the two fragment geometries to be fixed for this purpose.

The first step is to assign charges to each atom. Partial charges are assigned depending on the atomic environment. these partial charges are then summed to arrive at the resultant charge on the atom.

It turns out that these charges can have a large effect on the results of the energy minimization, and it is possible to modify them should you so wish.

The next step is to investigate the docking process. DTMM3 will calculate the interaction energy as you alter the relative positions of the two amino acids. In **vector docking**, we choose an atomic position in either fragment and then examine how the interaction energy varies as the two fragments approach along the vector defined by the difference in the coordinates. In **unconstrained docking** we simply let the package move the two fragments around in order to reach an energy minimum. the fragment geometries are usually kept constant. The screen grab below shows the result of our unconstrained docking calculation.

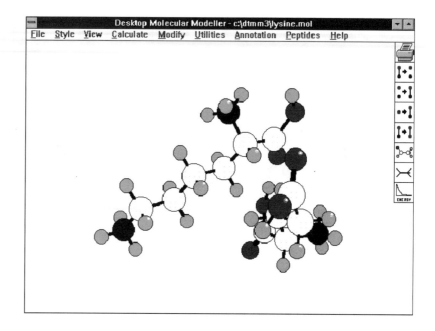

Figure 1.11   After optimization

2 The Hydrogen Molecule-ion and Potential Energy Surfaces

The traditional place to begin a quantum mechanical study of molecular structure is with the hydrogen molecule-ion H_2^+. Apart from being a prototype molecule, it gives a good place to begin our discussion of molecular electronic properties because molecules consist of **nuclei** in addition to the electrons, and we often have to be aware of the nuclear motion in order to understand the electronic ones. The two are linked.

In Chapter 1 I showed you how to model the vibrational motion of large molecules using molecular mechanics. In the MM method, we assume a simple potential for the nuclear motion, and then seek the minimum of the potential energy surface. Well, the particles generating the potential for this nuclear motion are the electrons. I hinted at this in the last chapter. A wealth of information, therefore, results from studies of molecular vibrational motions, and I will return again to this theme in a later chapter.

Central to our study is the concept of a **potential energy surface**, which you will meet time and time again in the following chapters. Let's start then with Figure 2.1. The H_2^+ molecule has an overall translational motion which I am going to ignore. For simplicity, I have drawn a local axis system with the centre of mass as the origin and by convention we call the internuclear axis, the z axis.

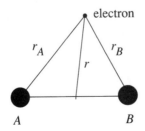

Figure 2.1   Hydrogen molecule-ion

We need to be clear about the various coordinates, and about the differences between various vector and scalar quantities. The electron has position vector $\mathbf{r}$, and the length of the vector is r. The scalar distance of nucleus A from the electron is r_A, r_B being the scalar distance between nucleus B and the electron. I will use R_{AB} to denote the scalar distance between nuclei A and B. The position vector for nucleus A is $\underline{\mathbf{R}}_A$ and the position vector for nucleus B is $\underline{\mathbf{R}}_B$. The wavefunction for the motion of the molecule as a whole (but ignoring the translational motion) will therefore depend on the vector quantities $\underline{\mathbf{R}}_A$, $\underline{\mathbf{R}}_B$ and $\mathbf{r}$

It is an easy step to write down the Hamiltonian operator for the problem

$$H_{\text{tot}} = -\frac{h^2}{8\pi^2 M}\nabla_A^2 - \frac{h^2}{8\pi^2 M}\nabla_B^2 + \frac{e^2}{4\pi\epsilon_0 R_{AB}} - \frac{h^2}{8\pi^2 m}\nabla^2 - \frac{e^2}{4\pi\epsilon_0 r_A} - \frac{e^2}{4\pi\epsilon_0 r_B}$$

The first two terms represent the kinetic energy of the nuclei A and B (each of mass M), whilst the fourth term represents the kinetic energy of the electron (of mass m). The fifth and sixth (negative) terms give the coulomb attraction between the nuclei and the electron. The third term is the coulomb repulsion between the nuclei. I have used the subscript 'tot' to mean nuclear plus electron.

As is often the case in quantum chemistry, we want to solve the time-independent Schrödinger equation

$$H_{\text{tot}}\Psi_{\text{tot}}(\underline{\mathbf{R}}_A, \underline{\mathbf{R}}_B, \mathbf{r}) = \epsilon_{\text{tot}}\Psi_{\text{tot}}(\underline{\mathbf{R}}_A, \underline{\mathbf{R}}_B, \mathbf{r})$$

and naturally the total wavefunction depends on the position vectors of both the nuclei and on the position vector of the electron.

2.1   THE BORN–OPPENHEIMER APPROXIMATION

The first thing to note is that the masses of the nuclei are very much greater than the mass of the electron. If they were classical particles, we would argue that their velocities would be very much less than the velocity of the electron, and so that to a first approximation the motion of the electron should be the same as if the nuclei were fixed in space.

If the motions of the electron and the two nuclei are indeed independent of one another, the total wavefunction should be a product of an electronic one and a nuclear one, $\Psi_{\text{tot}} = \psi_e\psi_n$. We might reasonably expect that the electronic wavefunction would depend on the particular values of R_A and R_B at which the nuclei were fixed.

The problem was tackled quantum-mechanically in 1927 by Born and Oppenheimer; their treatment is pretty involved, but the basic physical idea is as outlined above. To make the notation simpler, we rewrite H_{tot} as follows

$$H_{\text{tot}} = [-\frac{h^2}{8\pi^2 M}\nabla_A^2 - \frac{h^2}{8\pi^2 M}\nabla_B^2 + \frac{e^2}{4\pi\epsilon_0 r_{AB}}] - \frac{h^2}{8\pi^2 m}\nabla^2 - \frac{e^2}{4\pi\epsilon_0 r_A} - \frac{e^2}{4\pi\epsilon_0 r_B}$$

$$= H_n + H_e$$

where I have put H_n for [...], the **nuclear** terms in H_{tot}, and H_e for any remaining terms which involve the electrons. The nuclear part only depends on the coordinates of the nuclei, but the electronic part depends on the electrons **and** the nuclear positions (through the r_A and r_B).

If we had been dealing with a polyelectronic molecule, the electron–electron repulsion terms would have gone into H_e. We thus wish to solve

$$H_{tot}\Psi_{tot} = \epsilon_{tot}\Psi_{tot}$$
$$(H_n + H_e)\Psi_{tot} = \epsilon_{tot}\Psi_{tot}$$

If the motion of the nuclei and of the electrons were independent, the total wavefunction would be a product of a nuclear and an electronic one; $\Psi_{tot} = \psi_e\psi_n$ where the two wavefunctions ψ_e and ψ_n would be separately solutions of the 'nuclear' and 'electronic' equations $H_n\psi_n = \epsilon_n\psi_n$ and $H_e\psi_e = \epsilon_e\psi_e$. We need to investigate the conditions under which this is true, and to do this we use a technique called 'separation of variables'. Substitute $\Psi_{tot} = \psi_e\psi_n$ into the full equation to give

$$H_{tot}\Psi_{tot} = \epsilon_{tot}\Psi_{tot}$$
$$(H_n + H_e)\Psi_{tot} = \epsilon_{tot}\Psi_{tot}$$
$$(H_n + H_e)\psi_e\psi_n = \epsilon_{tot}\psi_e\psi_n$$

If we divide right and left-hand sides by $\Psi_{tot} = \psi_e\psi_n$ we get

$$\frac{1}{\psi_e\psi_n}(H_n + H_e)\psi_e\psi_n = \epsilon_{tot}$$

and we investigate whether the left hand side really is a constant.

Now, H_n contains differentials with respect to the nuclear coordinates; H_e contains differentials with respect to the electrons. The nuclear wavefunction ψ_n does not involve the coordinates of the electron, but the electronic wavefunction **does** involve the coordinates of the nuclei. So that $\nabla^2\psi_e\psi_n = \psi_n\nabla^2\psi_e$ but $\nabla_A^2\psi_e\psi_n = \psi_e\nabla_A^2\psi_n +$ extra terms. Born and Oppenheimer (1927) showed that, to a very good approximation, these extra terms were of the order of m/M_A (1 part in 2000) and so the motions of the nuclei and the electrons could indeed be considered separately.

There **are** phenomena such as the Renner effect where the Born–Oppenheimer approximation is not valid, but for 95% of chemical applications the Born–Oppenheimer approximation is a vital one. It has great conceptual importance; without it, we could not speak of a molecular geometry!

The electronic wavefunction is thus given by solution of $H_e\psi_e = \epsilon_e\psi_e$ and the total energy is

$$\epsilon_{tot} = \epsilon_{el} + \frac{e^2}{4\pi\epsilon_0 R_{AB}}$$

where the nuclei are to be thought of as being 'clamped' in position for the purpose of calculation of the electronic energy and electronic wavefunction. We then change the internuclear separation and recalculate the electronic problem at the new *A–B* distance. The nuclear motion is found by solving the relevant nuclear Schrödinger equation. Don't confuse this with the molecular mechanics technique discussed in Chapter 1. The nuclear equation is a quantum mechanical equation, and has to be solved by the standard techniques. You might like to read Eyring, Walter and Kimball (EWK)'s classic text *Quantum Chemistry*, to see how it is done. Often we substitute a model potential such as the Morse potential into this nuclear Schrödinger equation, in order to make the solution easier. That is a theme for later chapters.

Because the nuclei move in a potential energy due to the electrons, we talk about a **potential energy curve**.

The Born–Oppenheimer approximation shows us the way ahead for a polyelectronic molecule consisting of *n* electrons and *NUC* nuclei; we are generally interested in solving the electronic Schrödinger equation

$$H_e \Psi_e(\mathbf{r}_1, \mathbf{r}_2, \ldots, \mathbf{r}_n) = \epsilon_e \Psi_e(\mathbf{r}_1, \mathbf{r}_2, \ldots, \mathbf{r}_n)$$

at some nuclear geometry, with the *NUC* nuclei fixed at some positions in space. We then have to solve the nuclear problem with the nuclei moving in the potential due to the electrons if we wish to study the nuclear motion. More often than not, we will be content with the electronic problem alone.

You will find details of the solution of the electronic Schrödinger equation for H_2^+ in any standard quantum mechanics text (such as EWK) together with the potential energy curve (a plot of the total energy against bond distance). If you are particularly interested in the method of solution, the key reference is Bates, Lodsham and Stewart (1953). Even for such a simple case, it isn't easy! The problem has to be solved numerically. Burrau (1927) introduced the coordinates $r_a \pm r_b$, and was able to separate and integrate the electronic Schrödinger equation by direct numerical methods. The best solution for the ground state is that of Wind (1965) who found an electronic dissociation energy of $-0.6026342 E_h$. Very little direct experimental data is available for H_2^+, since the ground electronic state is the only one which is bound.

2.2   THE LCAO MODEL

An **exact** solution of the electronic Schrödinger equation is no mean feat, even for such a simple molecule as H_2^+. The electronic Schrödinger equation is

$$\left(-\frac{h^2}{8\pi^2 m}\nabla^2 - \frac{e^2}{4\pi\epsilon_0 r_A} - \frac{e^2}{4\pi\epsilon_0 r_B}\right)\psi_e = \epsilon_e \Psi_e$$

Let's examine the limiting behaviour when the electron is in the vicinity of nucleus *A* but far away from nucleus *B*. In this case the electronic Schrödinger equation becomes

$$\left(-\frac{h^2}{8\pi^2 m}\nabla^2 - \frac{e^2}{4\pi\epsilon_0 r_A}\right)\Psi_e = \epsilon_e \Psi_e$$

which is just a hydrogen atom problem. So, near each nucleus, the molecular H_2^+ wavefunction will resemble an atomic orbital centred on that particular nucleus. This suggests that it would be profitable to investigate the approximate wavefunction

$$\Psi_{LCAO} = c_A 1s_A + c_B 1s_B$$

where $1s_A$ represents a hydrogen 1s orbital centred on nucleus A, and $1s_B$ represents a 1s orbital centred on nucleus B. These atomic orbitals have the algebraic form (e.g.)

$$1s_A = \sqrt{\frac{\zeta^3}{\pi a_0^3}}\exp\left(-\zeta\frac{r_A}{a_0}\right)$$

and for a hydrogen atom the orbital exponent $\zeta = 1$. We refer to this treatment as the **Linear Combination of Atomic Orbitals (LCAO)** method.

Just by simple symmetry arguments we can deduce that $c_A = \pm c_B$, and we label the two LCAO-molecular orbitals by symmetry; $1\sigma_g = 1s_A + 1s_B$ and $1\sigma_u = 1s_A - 1s_B$. Neither is a solution of the electronic Schrödinger equation, but each has the correct boundary conditions and so is a possible approximate solution.

In order to test the accuracy of the LCAO approximations, we recall the **variation principle**; if Ψ_{LCAO} is an approximate solution then the variational integral

$$\epsilon_{LCAO} = \frac{\int \Psi_{LC4AO} H \Psi_{LCAO} d\tau}{\int \Psi_{LCAO}^2 d\tau}$$

gives an upper bound to the true energy. In this particular case, evaluation of the variational integral is reasonably simple, and after a little algebraic manipulation you will find that

$$\epsilon_{LCAO} = \frac{H_{AA} \pm H_{AB}}{1 \pm S_{AB}}$$

where the + sign goes with the $1s_A + 1s_B$ combination, the − sign with the $1s_A - 1s_B$ combination and the integrals are

$$H_{AA} = \int 1s_A H_e 1s_A d\tau$$

$$H_{AB} = \int 1s_A H_e 1s_B d\tau$$

$$S_{AB} = \int 1s_A 1s_B d\tau$$

2.3   INTEGRAL EVALUATION

One of the biggest headaches in computational quantum chemistry is the problem of integral evaluation, so let's spend a few minutes tackling this very simple problem. The physical quantities h, e and m_e all tend to get in the way, so the first task is to write the Hamiltonian in dimensionless form (each variable in the equation is now the true value divided by its atomic unit). I showed you how to do this in Chapter 0. The 'real' electronic Hamiltonian

$$H_{el} = -\frac{h^2}{8\pi^2 m}\nabla^2 - \frac{e^2}{4\pi\epsilon_0 r_A} - \frac{e^2}{4\pi\epsilon_0 r_B}$$

becomes a 'reduced' one

$$H_{el} = -\frac{1}{2}\nabla^2 - \frac{1}{r_A} - \frac{1}{r_B}$$

and the 'energy' we calculate has to be multiplied by the atomic unit of energy, E_h.

The variational energy expression (for the electronic energy) becomes

$$\epsilon_{el} = \frac{1}{1+S_{AB}}\left(\int 1s_A(-\frac{1}{2}\nabla^2)1s_A d\tau + \int 1s_A(-\frac{1}{2}\nabla^2)1s_B d\tau\right.$$
$$\left. + \int 1s_A(-\frac{1}{r_A})1s_A d\tau + \int 1s_A(-\frac{1}{r_B})1s_A d\tau + 2\int 1s_A(-\frac{1}{r_B})1s_B d\tau\right)$$

which I am going to write as

$$\epsilon_{el} = \frac{1}{1+S_{AB}}(T_{AA} + T_{AB} - V_{AARA} - V_{AARB} - 2V_{ABRA})$$

Let's take the overlap integral $S_{AB} = \int 1s_A(\mathbf{r})1s_B(\mathbf{r})d\tau$, where the integral has to be done over the three coordinates of the electron whose position vector is $\mathbf{r}$. For the sake of generality, I will take atomic orbitals with exponent ζ so that their normalizing factors are $(\zeta^3/\pi)^{1/2}$. The first step in integral evaluation is to choose an appropriate coordinate system. In this instance, it is the so-called elliptic system; $\mu = (r_A + r_B)/R_{AB}$, $\nu = (r_A - r_B)/R_{AB}$ and the angle ϕ which measures the rotation about the internuclear axis. The ranges for integration are $[1, \infty]$, $[-1, +1]$ and $[0, 2\pi]$ and the volume element $d\tau = (R_{AB}^3/8)(\mu^2 - \nu^2)d\mu d\nu d\phi$. You might care to try evaluation of these integrals. See EWK if this kind of thing interests you. For the sake of completeness, I have summarized all the $H_2{}^+$ integrals in Table 2.1.

It is a simple matter to calculate the energy using a spreadsheet, and the resulting potential energy curve for the + combination is reproduced (from Chapter 1) in Figure 2.2. I produced the graph using JANDEL's Tablecurve2D package for Windows.

Table 2.1 Diatomic integrals for the H_2^+ problem. R_{AB} = internuclear distance, $1s_A$ and $1s_B$ are the Slater orbitals on the nuclear centres each with exponent ζ. $p = \zeta R_{AB}$. Atomic units are used throughout.

S_{AA}	1
S_{AB}	$e^{-p}(1 + p + p^2/3)$
T_{AA}	$\frac{1}{2}\zeta^2$
T_{AB}	$\zeta^2 e^{-p}[\frac{1}{2} + p/2 - (1/6)p^2]$
V_{AARA}	ζ
V_{ABRA}	$\zeta e^{-p}(1 + p)$
V_{AARB}	$(1/R_{AB})[1 - e^{-2p}(1 + p)]$

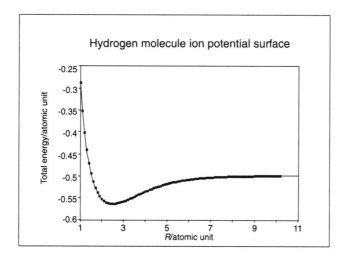

Figure 2.2 Potential energy surface for H_2^+

At large internuclear separation, the $1s_A + 1s_B$ combination gives a good description of the dissociation of H_2^+ into $H + H^+$ (which have total energy $-\frac{1}{2}E_h$). The only stable electronic state of H_2^+ is the lowest $^1\Sigma_g$ state, and electronic band spectra indicate a dissociation energy of $D_e = 2.791$ eV ($0.01026E_h$) and equilibrium bond length $R_e = 106$ pm ($2.003a_0$). Our simple LCAO model predicts $0.00647E_h$ and $2.495a_0$, in qualitative agreement with experiment. A more detailed comparison with the 'true' potential energy curve shows that the $1s_A + 1s_B$ LCAO combination gives poor agreement with experimental results for small internuclear separations. One reason is that, at small R_{AB}, the limiting model should really be a helium cation He^+ for which the 1s atomic orbital has an exponent of 2 rather than 1.

2.4   IMPROVING THE ATOMIC ORBITAL

With this in mind, it is sensible to modify the atomic 1s orbital by treating the orbital exponent ζ as a variational parameter. What we do is to vary ζ for each value of the internuclear separation, and for each value of R we calculate the energy with that particular orbital exponent. Just for illustration, I calculated the energy for a range of values of the orbital exponent and internuclear separation, and the results are shown in the contour diagram below. The x axis corresponds to internuclear separations running from $1.8a_0$ to $2.2a_0$ and the vertical axis corresponds to ζ values running from 1 to 1.5. It turns out that, around the potential energy minimum, $\zeta = 1.24$. We say that the 'atomic orbital contracts on molecule formation'.

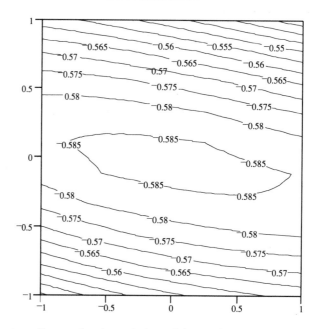

Figure 2.3   Contour diagram for the variation of the total energy ϵ with internuclear separation (the x axis) and orbital exponent (the y axis)

Table 2.2   Historically significant calculations for H_2^+ ground state

Description	Reference	D_e (eV)	Re (pm)
experiment		2.791	106.0
numerical	Wind	2.792	105.8
single hydrogenic 1s	$\zeta_{1s} = 1$	1.763	132.3
best 1s exponent	Finkelstein and Horowitz (1928) $\zeta_{1s} = 1.238$	2.354	106.8
simple LCAO 1s and $2p_\sigma$ with same exponent	Dickinson (1933) $\zeta = 1.254$ ratio $p/s = 0.1605$	2.716	106.1

2.5 THE LCAO APPROACH

Why stop with a single hydrogenic 1s atomic orbital on either centre? A little thought shows that (e.g.) $2p_\sigma$ atomic orbitals should also be important in describing the bonding and so we could write

$$\psi_{\text{trial}} = c_A 1s_A + c_B 1s_B + d_A 2p_{\sigma,A} + d_B 2p_{\sigma,B}$$

It is obvious by symmetry that $c_A = \pm c_B$ and $d_A = \pm d_B$, but what about the ratio of the values of c to d? I'll just mention for now that there is a systematic procedure called the **Hartree–Fock Self Consistent Field method** for solving this problem. In the special case of H_2^+, which has only a single electron, we can just calculate the variational integral and require it to be the lowest possible value. Dickinson (1933) first did the calculation, and he also found the best values of the two orbital exponents at the internuclear equilibrium distance to be $\zeta_{1s} = 1.246$ and $\zeta_{2p} = 2.965$ (see Table 2.2).

A deal of emphasis is placed on the **visualization** of quantum chemical calculations. I emphasized the importance of graphical user interfaces in Chapter 1. There are several ways of representing molecular orbitals graphically. Figures 2.4–2.6 all refer to the simple diatomic LCAO treatment of the electronic ground state of H_2^+

$$\psi_{\text{LCAO}} = \frac{1}{\sqrt{2(1 + S_{AB})}} (1s_A + 1s_B)$$

The bond length was taken to be 106 pm ($= 2.003 a_0$), and the orbital exponent $\zeta = 1$. First of all, it is traditional to plot values of ψ along the internuclear axis, and this is shown in Figure 2.4.

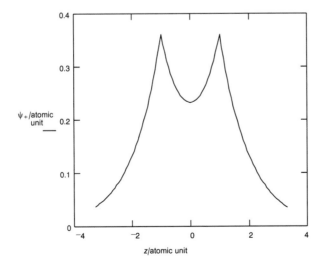

Figure 2.4 H_2^+ LCAO treatment. ψ versus distance along the internuclear axis

As an alternative, we might want to plot ψ^2 instead of ψ on the physical grounds that such quantities have a direct physical interpretation (the charge density). The form of the graph is very similar to Figure 2.4, and I haven't shown it.

Many graphics packages are much more sophisticated, and we can easily represent the MO or its square as a contour diagram in the molecular plane (Figure 2.5)

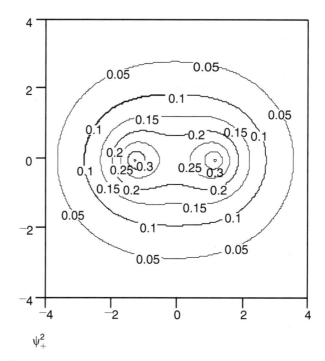

ψ_+^2

Figure 2.5   Molecular orbital density contours (atomic units)

or alternatively as a surface plot (Figure 2.6). I made all these diagrams quite simply on my office PC using Mathcad5.0.

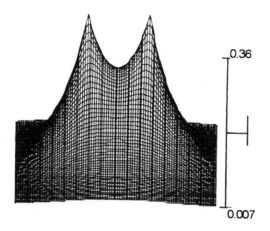

0.36

0.007

Figure 2.6   Surface plot for the $1\sigma_g$ orbital in H_2^+ (LCAO calc)

3 The Hydrogen Molecule

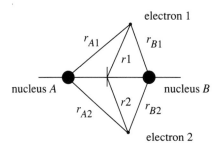

Figure 3.1 The hydrogen molecule

I dealt with the simplest possible molecule, H_2^+, in Chapter 2. The hydrogen molecule is also one of great historical interest, and it is shown schematically in Figure 3.1. I have used the same naming conventions as for H_2^+, and ignored the overall motion of the molecule. The coordinate origin is half way along the H-H bond.

To solve the time independent Schrödinger equation for the motion (of the nuclei + electrons, but neglecting the overall translation of the molecule) we have to start from the Hamiltonian operator

$$H_{tot} = [-\frac{h^2}{8\pi^2 M}\nabla_A^2 - \frac{h^2}{8\pi^2 M}\nabla_B^2 + \frac{e^2}{4\pi\epsilon_0 R_{AB}}] - \frac{h^2}{8\pi^2 m}\nabla_1^2 - \frac{h^2}{8\pi^2 m}\nabla_2^2$$

$$-\frac{e^2}{4\pi\epsilon_0 r_{A1}} - \frac{e^2}{4\pi\epsilon_0 r_{B1}} - \frac{e^2}{4\pi\epsilon_0 r_{A2}} - \frac{e^2}{4\pi\epsilon_0 r_{B2}} + \frac{e^2}{4\pi\epsilon_0 r_{12}}$$

just as for H_2^+.

The terms in square brackets are to do with the nuclear motion; the first two represent the kinetic energy of the nuclei labelled A and B (of mass M), and the third term in the square brackets is the coulomb repulsion between the two nuclei. The fourth and fifth terms give the kinetic energy of the two electrons. The next four negative terms give the mutual coulomb attraction between the two nuclei A,B and the two electrons labelled $1,2$. The final term is the coulomb repulsion between electron 1 and 2, with r_{12} the distance between them. As in Chapter 1, I have used the subscript 'tot' to mean nuclear plus electron. The first step is to make use of the Born–Oppenheimer approximation, so I separate the nuclear and the electronic terms

$$H_{\text{tot}} = H_{\text{n}} + H_{\text{e}}$$

where H_{n} is the terms in square brackets. Note that the 'electronic' contribution H_{e} contains terms that refer to electron 1, terms that refer to electron 2 and a cross term involving r_{12}.

For large molecules, very many similar terms contribute to the Hamiltonian. To simplify the notation I am going to collect together all those terms that depend explicitly on the coordinates of each electron (e.g., electron 1), and write them as

$$h(\underline{r}_1) = -\frac{h^2}{8\pi^2 m}\nabla_1^2 - \frac{e^2}{4\pi\epsilon_0 r_{A1}} - \frac{e^2}{4\pi\epsilon_0 r_{B1}}$$

Such operators which collect together all the variable terms involving a particular electron (in this case electron 1) are called **one-electron operators**. The $1/r_{12}$ term is a typical **two-electron operator**, and we usually write it as

$$g(\underline{r}_1,\underline{r}_2) = \frac{e^2}{4\pi\epsilon_0 r_{12}}$$

It represents the coulomb repulsion between the two electrons.

You have probably come across the idea of electron spin, know that it can be represented by a single variable which we write as s, and are probably wondering when spin is going to appear in the discussion? Well, bear with me for a little while yet.

Using the notation of one and two-electron operators, the electronic Hamiltonian is

$$H_{\text{e}} = h(\underline{r}_1) + h(\underline{r}_2) + g(\underline{r}_1,\underline{r}_2)$$

The total wavefunction will depend on the spatial coordinates $\underline{r}_1$ and $\underline{r}_2$ of the two electrons 1,2 and also the spatial coordinates $\underline{R}_A$ and $\underline{R}_B$ of the two nuclei A, B. I will write this as $\Psi_{\text{tot}}(\underline{R}_A,\underline{R}_B,\underline{r}_1,\underline{r}_2)$. The Schrödinger equation is

$$H_{\text{tot}}\Psi_{\text{tot}}(\underline{R}_A,\underline{R}_B,\underline{r}_1,\underline{r}_2) = \epsilon_{\text{tot}}\Psi_{\text{tot}}(\underline{R}_A,\underline{R}_B,\underline{r}_1,\underline{r}_2)$$

but we can carry forward our knowledge from Chapter 1 about the Born–Oppenheimer equation and focus attention on the **electronic** problem. Thus we have to solve

$$H_{\text{e}}\Psi_e(\underline{r}_1,\underline{r}_2) = \epsilon_e\Psi_e(\underline{r}_1,\underline{r}_2)$$

or, in terms of our new notation

$$[h(\mathbf{r}_1) + h(\mathbf{r}_2) + g(\mathbf{r}_1, \mathbf{r}_2)]\Psi_e(\mathbf{r}_1, \mathbf{r}_2) = \epsilon_e \Psi_e(\mathbf{r}_1, \mathbf{r}_2)$$

As before, the nuclei are to be thought of as being clamped in position for the purpose of calculating the electronic energy and wavefunction.

If we want to calculate a potential energy surface, then we will have to change the internuclear separation and rework the electronic problem at various new $A–B$ distances, as in the $H_2{}^+$ calculation. And once again, should we be so interested, the nuclear motion can be found by solving the nuclear Schrödinger equation. This is a quantum mechanical equation, not to be confused with molecular mechanics.

You will see shortly that an **exact** solution of the electronic Schrödinger equation is impossible, because of the electron repulsion term $g(\mathbf{r}_1, \mathbf{r}_2)$. What we have to do is to investigate approximate solutions based on chemical intuition, and then refine these models, typically using the variation principle, until we attain the required accuracy. This means in particular that any approximate solution will **not** necessarily satisfy the electronic Schrödinger equation, and we will not be able to calculate the energy from an eigenvalue equation. First of all, let's see why the problem is so difficult.

3.1 THE NON-INTERACTING ELECTRON MODEL

Imagine a model hydrogen molecule with non-interacting electrons, such that the electron repulsion is zero. Each electron still has a kinetic energy and is still attracted to both nuclei, but the electron motions are now completely independent of each other because the electron–electron interaction term is zero. We would, therefore, expect that the electronic wavefunction for the pair of electrons would be a product of the wavefunctions for two independent electrons in $H_2{}^+$ (Figure 3.2), which I will write $X(\mathbf{r}_1)$ and $Y(\mathbf{r}_2)$. Thus, $X(\mathbf{r}_1)$ and $Y(\mathbf{r}_2)$ are molecular orbitals which describe the independent motions of the two electrons in our non-interacting electron model H_2.

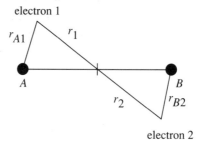

Figure 3.2 H_2 molecule with hypothetical independent electrons

In order to investigate whether the wavefunction can indeed be written in this way, we can use the separation of variables technique and so try a solution of the form

$$\Psi_e(\mathbf{r}_1, \mathbf{r}_2) = X(\mathbf{r}_1)\, Y(\mathbf{r}_2)$$

which we then substitute into the electronic Schrödinger equation. Divide each side of the equation by $\Psi_e(\mathbf{r}_1, \mathbf{r}_2)$ and you will find

$$\frac{1}{X(\mathbf{r}_1)} h(\mathbf{r}_1) X(\mathbf{r}_1) + \frac{1}{Y(\mathbf{r}_2)} h(\mathbf{r}_2) Y(\mathbf{r}_2) = \epsilon_e$$

Each term on the left-hand side separately involves the coordinates of the two electrons, and as the sum has to be a constant for all values of the coordinates of these two electrons, the terms must individually be constants which I can call ϵ_x and ϵ_y so

$$h(\mathbf{r}_1) X(\mathbf{r}_1) = \epsilon_x X(\mathbf{r}_1)$$
$$h(\mathbf{r}_2) Y(\mathbf{r}_2) = \epsilon_y Y(\mathbf{r}_2)$$

and in the case where the two electrons don't interact with each other we just solve the $H_2{}^+$ problem twice over (once for each electron). The solution I have given is exact, even for two atoms at a chemical bond length. If you substitute the solution into the electronic eigenvalue problem then you will find that it fits exactly, and you can calculate the energy that way if you so wish (or just add ϵ_x and ϵ_y).

You are probably used to this idea from descriptive chemistry, where we build up the configurations for many-electron atoms in terms of atomic wavefunctions, and where we would write an electronic configuration for He as $(1s)^2(2s)^2(2p)^6$. Unfortunately, the electron repulsion term $g(\mathbf{r}_1, \mathbf{r}_2)$ is not negligible. Even if it were, we would have to solve the electronic Schrödinger equation appropriate to naphthalene[67+] in order to make progress with the solution of the relevant electronic Schrödinger equation for naphthalene. Every molecular problem would be different. We therefore abandon the idea of non-interacting electrons and think again.

We will actually use the idea that the interaction between electrons can be **averaged**, in a later chapter; it forms the idea for the self consistent field model.

3.2 THE VALENCE BOND METHOD

So, let's get a bit more chemical and imagine the formation of a H_2 molecule from two separated H atoms, H_A and H_B initially an infinite distance apart. Electron 1 is associated with nucleus A, electron 2 with nucleus B and so the terms in the electronic Hamiltonian involving r_{AB}, r_{A2} and r_{B1} are all negligible when the nuclei are at infinity. Thus the electronic Schrödinger equation becomes

$$H_e \Psi_e(\mathbf{r}_1, \mathbf{r}_2) = (-\frac{h^2}{8\pi^2 m} \nabla_1^2 - \frac{h^2}{8\pi^2 m} \nabla_2^2$$
$$- \frac{e^2}{4\pi\epsilon_0 r_{A1}} - \frac{e^2}{4\pi\epsilon_0 r_{B2}}) \Psi_e(\mathbf{r}_1, \mathbf{r}_2)$$

and I am going to leave you to prove for yourself that the wavefunction corresponding to the solution of this 'infinite distance H_2' problem is a product of hydrogen atom wavefunctions. Physically, you should have expected this; the two atoms are independent and so the electronic wavefunctions **multiply** to give the molecular electronic wavefunction.

Let's write possible atomic orbitals for hydrogen atom A as χ_A and possible atomic orbitals for hydrogen atom B as χ_B, so that the molecular electronic wavefunction will be given as $\Psi_e(\mathbf{r}_1, \mathbf{r}_2) = \chi_A(\mathbf{r}_1)\chi_B(\mathbf{r}_2)$. The χ values can each be 1s, 2s, 2p ... hydrogen atomic orbitals. The lowest energy solution will be when the χ's correspond to 1s orbitals on the two hydrogen atoms, the next highest energy solutions will be when the χ's are 1s and 2s orbitals and so on. Possible solutions of the electronic problem, for the two H atoms at infinity, are shown in Table 3.1. However, there are a couple of loose ends.

Table 3.1 Hydrogen molecule with infinite atomic separation: electron 1 on nucleus A, electron 2 on nucleus B

Electronic wavefunction $\Psi_e(\mathbf{r}_1, \mathbf{r}_2)$	Electronic energy
$1s_A(\mathbf{r}_1)1s_B(\mathbf{r}_2)$	$2\epsilon(H, 1s)$
$1s_A(\mathbf{r}_1)2s_B(\mathbf{r}_2)$	$\epsilon(H, 1s) + \epsilon(H, 2s)$
$2s_A(\mathbf{r}_1)1s_B(\mathbf{r}_2)$	$\epsilon(H, 1s) + \epsilon(H, 2s)$
$2s_A(\mathbf{r}_1)2s_B(\mathbf{r}_2)$	$2\epsilon(H, 2s)$

3.3 INDISTINGUISHABILITY

I live in a country park, which has a lake, ducks and geese. Surely all ducks are the same? Certainly after the mating season the ducklings tend to all look alike, as shown, but of course we can distinguish between them; one likes bread and another doesn't and so on. Each duckling has its own characteristics and, in any case, we could if we wished paint one of them white, one blue . . . or give each of them a unique label. This might create a furore with the nature lovers, but despite the fact that ducklings all look the same, they are **distinguishable**. All similar things in our world are distinguishable.

Electrons are **indistinguishable**, they simply cannot be labelled. This means that an acceptable electronic wavefunction has to treat all electrons on an equal footing. Thus, although I have implied that electron 1 is somehow associated with nucleus H_A and electron 2 with nucleus H_B, I must also cater for the alternative description where electron 1 is associated with nucleus B and electron 2 with nucleus A.

In the case of infinite nuclear separation, the wavefunctions quoted above are exact solutions of the electronic Schrödinger equation and we can calculate their energies exactly. If we write $\epsilon(H, 1s)$ for a hydrogen 1s electron energy, $\epsilon(H, 2s)$ for a hydrogen 2s electron energy, etc., then Table 3.2 also gives energy solutions for the infinite distance H_2 problem.

Table 3.2 Hydrogen molecule with infinite atomic separation taking account of indistinguishability

Electronic wavefunction $\Psi_e(\mathbf{r}_1, \mathbf{r}_2)$		Electronic energy
$1s_A(\mathbf{r}_1)1s_B(\mathbf{r}_2)$	$1s_A(\mathbf{r}_2)1s_B(\mathbf{r}_1)$	$2\epsilon(H, 1s)$
$1s_A(\mathbf{r}_1)2s_B(\mathbf{r}_2)$	$1s_A(\mathbf{r}_2)2s_B(\mathbf{r}_1)$	$\epsilon(H, 1s) + \epsilon(H, 2s)$
$2s_A(\mathbf{r}_1)1s_B(\mathbf{r}_2)$	$2s_A(\mathbf{r}_2)1s_B(\mathbf{r}_1)$	$\epsilon(H, 1s) + \epsilon(H, 2s)$
$2s_A(\mathbf{r}_1)2s_B(\mathbf{r}_2)$	$2s_A(\mathbf{r}2)2s_B(\mathbf{r}_1)$	$2\epsilon(H, 2s)$

The extra terms (in the second column) have to be considered, as they are also allowed solutions. The ground state of the molecule is described equally by the two quantum states $1s_A(\mathbf{r}_1)1s_B(\mathbf{r}_2)$ and $1s_A(\mathbf{r}_2)1s_B(\mathbf{r}_1)$, and each quantum state has the same energy. Any linear combination of these two quantum states also has the same energy, so that we ought to consider a linear combination

$$a1s_A(\mathbf{r}_1)1s_B(\mathbf{r}_2) + b1s_A(\mathbf{r}_2)1s_B(\mathbf{r}_1)$$

Symmetry arguments tell us that we have to take $a = \pm b$.

3.4 ELECTRON SPIN

So far, I have ignored the fact that electrons have spin, which is a two-valued quantity. We write the two possible single electron spin states α and β, and the spin variable is denoted by s. The notation I am going to use is that $\alpha(s_1)$ means electron 1 in spin state α. For two electrons, we have to consider the possible combinations

$$\alpha(s_1)\beta(s_2) \pm \alpha(s_2)\beta(s_1)$$
$$\alpha(s_1)\alpha(s_2)$$
$$\beta(s_1)\beta(s_2)$$

3.5 THE PAULI PRINCIPLE

The total electronic wavefunction has to be constructed as a product of the spatial part and the spin part. For the electronic ground state we have to consider combinations of **spatial** functions

$$1s_A(\underline{r}_1)1s_B(\underline{r}_2) \pm 1s_A(\underline{r}_2)1s_B(\underline{r}_1)$$

with the four **spin** functions listed above. The generalized Pauli principle (1925) guides us in our choice; acceptable electronic wavefunctions have to be **antisymmetric** to the exchange of electron names, and so only the following combinations are physically acceptable:

$$[1s_A(\underline{r}_1)1s_B(\underline{r}_2) + 1s_A(\underline{r}_2)1s_B(\underline{r}_1)][\alpha(s_1)\beta(s_2) - \alpha(s_2)\beta(s_1)]$$

$$[1s_A(\underline{r}_1)1s_B(\underline{r}_2) - 1s_A(\underline{r}_2)1s_B(\underline{r}_1)][\alpha(s_1)\beta(s_2) + \alpha(s_2)\beta(s_1)]$$
$$[1s_A(\underline{r}_1)1s_B(\underline{r}_2) - 1s_A(\underline{r}_2)1s_B(\underline{r}_1)]\alpha(s_1)\alpha(s_2)$$
$$[1s_A(\underline{r}_1)1s_B(\underline{r}_2) - 1s_A(\underline{r}_2)1s_B(\underline{r}_1))\alpha(s_1)\beta(s_2)$$

In the infinite atom approximation, all four electronic wavefunctions have the same energy. But they correspond to different eigenvalues of the spin operators; the first wavefunction describes the ground singlet state of H_2 and the following three wavefunctions describe the three spin components of the lowest energy triplet spin state.

3.6 THE H_2 MOLECULE

For the real-life H_2 molecule, we cannot ignore the electron repulsion term $g(\underline{r}_1, \underline{r}_2)$ in the Hamiltonian. The simple atomic orbital product wavefunctions in Table 3.2 are no longer solutions of the electronic problem, but they have all the right physical characteristics and so are acceptable models. As the two H atoms approach, the energy will decrease. As the two H atoms get very close together, we expect that the energy will increase rapidly, because the hydrogen nuclei get close together and repel each other. Heitler and London first tried this approach in 1927 (Heitler and London 1927); in order to calculate the energy for their model wavefunction Ψ_{HL} we work out the variational integral

$$\epsilon_{HL} = \frac{\int \Psi_{HL} H \Psi_{HL} d\tau}{\int \Psi_{HL}^2 d\tau}$$

where the integration $\int \dots d\tau$ is over 'all space', which means the coordinates of both electrons. Such integrals are therefore six-dimensional, and far from easy.

We defined certain integrals ϵ_{AA} and ϵ_{AB} in our discussion of H_2^+, and the (electronic part of the) energy of the Heitler–London function can be written in terms of them as

$$\epsilon_{HL} = 2\epsilon(H, 1s) + \frac{J_{AB} + K_{AB}}{1 + S_{AB}^2}$$

where the integrals J_{AB} and K_{AB} are related to ϵ_{AA} and ϵ_{AB} as follows

$$J_{AB} = -2\epsilon_{AA} + \int\int 1s_A^2(\mathbf{r}_1)g(\mathbf{r}_1,\mathbf{r}_2)1s_B^2(\mathbf{r}_2)d\tau_1 d\tau_2$$

$$K_{AB} = -2S\epsilon_{AB} + \int\int (1s_A(\mathbf{r}_1)1s_B(\mathbf{r}_2))^2 g(\mathbf{r}_1,\mathbf{r}_2)d\tau_1 d\tau_2$$

The integrals of the type $\dots \int\int d\tau_1 d\tau_2$ are called **two-electron integrals** and they are particularly hard to evaluate. They can be given classical electrostatic interpretations; the first one represents the mutual potential energy of two charge clouds with densities $-e1s_A^2(\mathbf{r}_1)$ and $-e1s_B^2(\mathbf{r}_2)$. The second one represents the mutual potential energy of two charge clouds, each of which has the same density $-e1s_A 1s_B$. We now have to remember the variation with bond distance, and calculate the variational integral for a range of R_{AB} (i.e., we calculate the potential energy curve)

$$\epsilon_{tot} = \epsilon_{HL} + \frac{e^2}{4\pi\epsilon_0 R_{AB}}$$

We often refer to Heitler and London's method as the **valence bond** (VB) method. A comparison between the experimental and the VB potential energy curves shows an excellent agreement at large R but poor in the valence region (Table 3.3). The cause of this lies in the method itself; the VB method starts from atomic wavefunctions and adds

Table 3.3 Significant Simple calculations for H_2

Description	Reference	D_e/eV	R_e/pm
Experiment		4.72	74.1
single hydrogenic 1s valence bond	Heitler and London $\zeta_{1s} = 1$	3.156	86.8
Best 1s exponent at same R	$\zeta_{1s} = 1.12$	3.592	
single hydrogenic 1s, VB configuration interaction	$\zeta_{1s} = 1$	3.238	88.4
Single hydrogenic 1s, LCAO-MO	$\zeta_{1s} = 1$	2.694	84.7
Best 1s exponent at same R	$\zeta_{1s} = 1.14$	3.292	
simple LCAO with 1s and $2p_\sigma$ AOs	Rosen (1931)	4.02	

as a perturbation the fact that the electron clouds of the atoms are polarized when the molecule is formed. A slight improvement in the predicted dissociation energy occurs if the 1s atomic orbital exponent is treated as a variational parameter and the bond distance kept constant.

3.7 CONFIGURATION INTERACTION

The simple VB treatment of H_2 uses a wavefunction $[1s_A(\underline{r}_1)1s_B(\underline{r}_2) + 1s_A(\underline{r}_2)1s_B(\underline{r}_1)]$ $[\alpha(s_1)\beta(s_2) - \alpha(s_2)\beta(s_1)]$ and it is usual to give a physical interpretation to each of the spatial terms; $1s_A(\underline{r}_1)1s_B(\underline{r}_2)$ represents a situation where electron 1 is associated with nucleus A and electron 2 with nucleus B. We talk about 'covalent structures', and recognise that 'ionic' structures such as $1s_A(\underline{r}_1)1s_A(\underline{r}_2)$ should also be considered. Thus, an improved VB wavefunction would be

$$\{a[1s_A(\underline{r}_1)1s_B(\underline{r}_2) + 1s_A(\underline{r}_2)1s_B(\underline{r}_1)]$$
$$+b[1s_A(\underline{r}_1)1s_A(\underline{r}_2) + 1s_B(\underline{r}_1)1s_B(\underline{r}_2)]\}[\alpha(s1)\beta(s2) - \alpha(s2)\beta(s1)]$$

Weinbaum used the variation method to find the values of a and b using both simple atomic 1s orbitals with exponent 1 and with optimized exponent. This is an example of **configuration interaction**.

3.8 THE LCAO MOLECULAR ORBITAL METHOD

We saw in chapter 2 how to find simple LCAO wavefunctions for $H_2{}^+$ in terms of hydrogenic atomic orbitals. Starting from a single 1s orbital on each centre, we constructed LCAO approximations to the two lowest energy molecular orbitals as $1\sigma_g = 1s_A + 1s_B$ and $1\sigma_u = 1s_A - 1s_B$, and we noted how to calculate their energies. We now have to deal with two electrons, and we need also to concern ourselves with spin and with antisymmetry. Table 3.4 summarizes possible allowed space and spin parts.

Table 3.4   Possible LCAO MO wavefunctions for H_2

Space part	Spin part	State	
		Symmetry	Configuration
$1\sigma_g(\underline{r}_1)1\sigma_g(\underline{r}_2)$	$\alpha(s_1)\beta(s_2) - \alpha(s_2)\beta(s_1)$	$^1\Sigma_g$	$1\sigma_g^2$
$1\sigma_g(\underline{r}_1)1\sigma_u(\underline{r}_2) - 1\sigma_u(\underline{r}_1)1\sigma_g(\underline{r}_2)$	$\alpha(s_1)\alpha(s_2)$ $\alpha(s_1)\beta(s_2) + \alpha(s_2)\beta(s_1)$ $\beta(s_1)\beta(s_2)$	$^3\Sigma_u$	$1\sigma_g^1 1\sigma_u^1$
$1\sigma_g(\underline{r}_1)1\sigma_u(\underline{r}_2)+$ $1\sigma_u(\underline{r}_1)1\sigma_g(\underline{r}_2)$	$\alpha(s_1)\beta(s_2) - \alpha(s_2)\beta(s_1)$	$^1\Sigma_u$	$1\sigma_g^1 1\sigma_u^1$
$1\sigma_u(\underline{r}_1)1\sigma_u(\underline{r}_2)$	$\alpha(s_1)\beta(s_2) - \alpha(s_2)\beta(s_1)$	$^1\Sigma_g$	$1\sigma_u^2$

Calculation of the potential energy curve for the first wavefunction gives rather poor agreement with experimental results for both the dissociation energy and for the bond distance.

3.8.1 Configuration interaction

In molecular orbital language, we write $1\sigma_g^2$ for the ground-state electronic configuration, $1\sigma_g^1 1\sigma_u^1$ for a singly excited state and $1\sigma_u^2$ for the first available double excited state. In the spirit of the variation technique, we would seek to improve the ground state wavefunction by writing a 'better' electronic wavefunction as

$$\Psi_{better} = a\Psi(1\sigma_g^2) + b\Psi(1\sigma_g^1 1\sigma_u^1) + c\Psi(1\sigma_u^2)$$

where a, b and c have to be determined from the variation principle. The symmetry of the singly excited state is different from that of the ground state, and so it does not contribute. Once the calculation is done, we do indeed find an improved bond dissociation energy and bond distance.

3.8.2 Computer program

This is once again configuration interaction at work, and the demonstration computer program 'H2MOL' on your disk should take all the hard work out of the calculation; the BASIC program will do all the calculations for the H_2 molecule with 1s-type orbitals on either centre. You might like to calculate the potential energy curves for the LCAO and the LCAO-CI treatments, with a fixed orbital exponent $\zeta = 1$. In which case the two graphs are given in Figures 3.3 and 3.4.

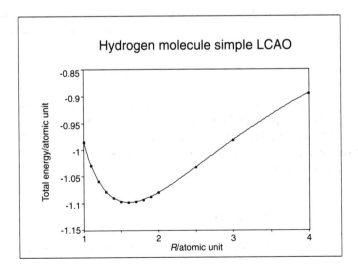

Figure 3.3   Potential energy curve for H_2 molecule, simple LCAO wavefunction

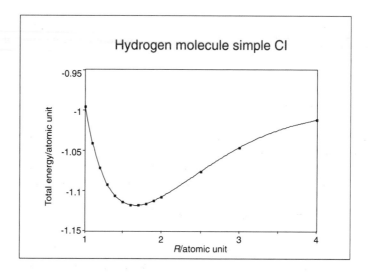

Figure 3.4 H_2 molecule simple LCAO calculation, CI treatment

Note that the simple LCAO treatment gives incorrect dissociation products; the dissociation products of the reaction $H_2 \rightarrow 2H(^2s)$ is $-1E_h$. The LCAO calculation gives $H_2 \rightarrow H^+ + H^-$. This incorrect treatment of dissociation products is a general failing of all simple LCAO treatments when a bond is reasonably covalent. The formulae for the one-electron integrals needed in the energy calculations have been given in Chapter 2. The two-electron integrals can be found in the work of Roothaan (1951), Suguira (1927) and Ruedenberg (1956).

3.9   COMPARISON OF LCAO-MO AND VB

It is instructive to compare this simple LCAO wavefunction with the simple valence bond function given above. To cut a long story short, the LCAO wavefunction can be written

$$\psi(\text{LCAO}) = \psi(\text{VB}) + \text{ionic structures.}$$

where equal weights are given to the 'covalent' and to the 'ionic' structures. The simple VB technique gives a weighting of zero to the ionic structures. In their simplest forms, the simple LCAO-MO and VB techniques appear to give a very different description of the dihydrogen covalent bond. At the end of the day, when CI is taken into account, both give exactly the same answers.

3.10   SLATER DETERMINANTS

The simplest possible LCAO building block which I have given you is

$$1\sigma_g(\underline{\mathbf{r}}_1)1\sigma_g(\underline{\mathbf{r}}_2)[\alpha(s_1)\beta(s_2) - \alpha(s_2)\beta(s_1)]$$

which takes account of spin and antisymmetry. This can be written as a determinant

$$\begin{vmatrix} 1\sigma_g(\underline{r}_1)\alpha(s_1) & 1\sigma_g(\underline{r}_2)\alpha(s_2) \\ 1\sigma_g(\underline{r}_1)\beta(s_1) & 1\sigma_g(\underline{r}_2)\beta(s_2) \end{vmatrix}$$

Determinants have the useful property that they change sign whenever we swop around two columns, which is equivalent to interchanging the names of the two electrons.

Such a determinant is called a **Slater determinant**, after Slater (1929) and you should realize that a

> Slater determinant = smallest logical
> building block for electronic wavefunctions.

I **don't** mean that such a wavefuntion is necessarily very accurate; you saw a minute ago that such a LCAO treatment of H_2 is actually rather poor. I mean that, **in principle**, a single determinant can have the correct spatial and spin symmetry to represent a given electronic state. To give a counter example, the first excited state of H_2 **cannot** be represented correctly as a single determinant. The best we can get away with is to write

$$\begin{vmatrix} 1\sigma_g(\underline{r}_1)\alpha(s_1) & 1\sigma_u(\underline{r}_2)\beta(s_2) \\ 1\sigma_g(\underline{r}_1)\beta(s_1) & 1\sigma_u(\underline{r}_2)\beta(s_2) \end{vmatrix} + \begin{vmatrix} 1\sigma_g(\underline{r}_1)\beta(s_1) & 1\sigma_u(\underline{r}_2)\alpha(s_2) \\ 1\sigma_g(\underline{r}_1)\beta(s_1) & 1\sigma_u(\underline{r}_2)\alpha(s_2) \end{vmatrix}$$

or at the very least, we have to take consideration of the pair of determinants when considering the excited state of the molecule. We cannot ignore one of them at the expense of the other.

So, we have learned that some electronic configurations can be described as a single Slater determinant, but others can only be described as a linear combination of such Slater determinants, even at the lowest level of accuracy.

4 The Electron Density

As computational facilities get better, electronic wavefunctions tend to become more and more complicated. A configuration interaction (CI) calculation on a medium sized molecule might be a linear combination of a million Slater determinants, and it is very easy to lose sight of the 'chemistry' and the 'chemical intuition', to say nothing of the visualization of the results. Such a wavefunction seems to give no simple physical picture of the electron distribution and how it determines the molecular properties, and so we must seek to find a way of extracting the information that is chemically useful.

In Chapter 2, I showed you how to write a simple LCAO wavefunction for the electronic ground state of H_2^+ where the **single** electron has position vector $\underline{r}$; the **spatial** part is

$$\psi(\underline{r}) = \frac{1}{\sqrt{2(1+S)}}[1s_A(\underline{r}) + 1s_B(\underline{r})]$$

where S is the overlap integral between the two atomic 1s orbitals, and the factor $1/\sqrt{(2(1+S)}$ is called the *normalization coefficient*.

The **total** wavefunction describing the electron is a product of a **spatial** part and a **spin** part, $\psi_+(\underline{r})$ times $\alpha(s)$ or $\beta(s)$; there are thus two different quantum states having the same spatial part $\psi_+(\underline{r})$. In the absence of a magnetic field, these both have the same energy, which is why I didn't stress electron spin in the earlier chapter. Some authors denote the 'total' variables for the electron as $\underline{x} = \underline{r}s$, and the total wavefunction as $\Psi(\underline{r}, s)$ or just $\Psi(\underline{x})$. I have used a capital Ψ here to emphasize that the wavefunction depends on space **and** spin. The Born interpretation of quantum mechanics tells us that $\Psi^2(\underline{r}, s) \, d\tau \, ds$ gives the chance of finding the electron in the spatial volume element $d\tau$ with spin between s and $s + ds$. Also, probabilities have to sum to 1, so that

$$\int \Psi^2(\underline{r}, s) \, d\tau \, ds = 1$$

where the integration has to be over the spatial variables and the spin variable. A short calculation will show you that the H_2^+ LCAO wavefunction does indeed satisfy this requirement.

Most physical properties such as the electrostatic potential, the dipole moment and so on, do not depend on electron spin and so we can ask a slightly different question; what is the chance that we will find the electron in a certain region of space $d\mathbf{r}$ irrespective of spin? The answer is that we **integrate** over the spin variable, and this gives us the **electron density.** So, to take one particular quantum state for H_2^+

$$\Psi_+(\mathbf{r})\alpha(s) = \frac{1}{\sqrt{2(1+S)}}[1s_A(\mathbf{r}) + 1s_B(\mathbf{r})]\alpha(s)$$

then $\int \Psi^2(\mathbf{r})\alpha^2(s)\,ds$ gives the chance of finding the electron in $d\tau$, irrespective of spin. A short calculation shows that this is

$$\frac{1}{S(1+S)}\left[1s_A^2(\mathbf{r}) + 1s_B^2(\mathbf{r}) + 2 \times 1s_A(\mathbf{r})1s_B(\mathbf{r})\right]$$

which is the electron density mentioned above. The electron density is often given the special symbol $P_1(\mathbf{r})$, where the subscript 1 means 'single electron'.

What happens with a many-electron wavefunction (such as the one below, which relates to the simple VB treatment of H_2)?

$$\Psi_{VB}(\mathbf{r}_1,\mathbf{r}_2) = \frac{1}{\sqrt{2(1+S^2)}}[1s_A(\mathbf{r}_1)1s_B(\mathbf{r}_2) + 1s_A(\mathbf{r}_2)1s_B(\mathbf{r}_1)]$$

As noted above, the common molecular properties don't depend on electron spin. I have therefore averaged over the effect of electron spin by integrating it out. The Born interpretation of quantum mechanics tells us that $[\Psi_{VB}^2(\mathbf{r}_1,\mathbf{r}_2)]\,d\tau_1\,d\tau_2$ is the probability of finding electron 1 in $d\tau_1$ and simultaneously finding electron 2 in $d\tau_2$.

The quantity in square brackets $[\ldots]$ comes to

$$[\ldots] = \frac{1}{2(1+S^2)}\left[1s_A^2(\mathbf{r}_1)1s_B^2(\mathbf{r}_2) + 1s_A^2(\mathbf{r}_2)1s_B^2(\mathbf{r}_1)\right.$$
$$+ 1s_A(\mathbf{r}_1)1s_B(\mathbf{r}_2)1s_A(\mathbf{r}_2)1s_B(\mathbf{r}_1)$$
$$\left.+1s_A(\mathbf{r}_2)1s_B(\mathbf{r}_1)1s_A(\mathbf{r}_1)1s_B(\mathbf{r}_2)\right]$$

Apart from electron spin, many simple molecular electronic properties depend only on the probability of finding **either** electron in a region of space $d\tau$. The region $d\tau$ is to be regarded as a fixed region of space that could be from time to time occupied by any of the electrons. To find this from the expression above, we focus attention on one of the electrons (say 1) and then average over the coordinates of the remaining ones. In this case, we average over the coordinates of electron 2 since it is the only other one. In the simple VB case we calculate $\int \Psi_{VB}^2(\mathbf{r}_1,\mathbf{r}_2)\,d\tau_2$ which works out as

$$\frac{1}{2(1+S^2)}\left[1s_A^2(\mathbf{r}_1) + 1s_B^2(\mathbf{r}_1) + 1s_A(\mathbf{r}_1) \times S \times 1s_B(\mathbf{r}_1) + S \times 1s_B(\mathbf{r}_1)1s_A(\mathbf{r}_1)\right]$$

Electrons are indistinguishable, and so this probability is exactly equal to the chance of finding electron 2 in $d\tau_1$. The chance of finding **an** electron, **either** electron 1 or 2 in $d\tau_1$ is therefore twice the above expression.

We define the electron density to be $P_1(\mathbf{r}_1) = 2\int \Psi_{\text{LCAO}}^2(\mathbf{r}_1,\mathbf{r}_2)\,d\tau_2$, and usually write it without the subscript '1' on the $\mathbf{r}$ in order to emphasize that we are discussing the occupation of a certain volume of space $d\tau$ by any electron. You will find

$$P_1(\mathbf{r}) = \frac{1}{(1+S^2)}\left[1s_A^2(\mathbf{r}) + 1s_B^2(\mathbf{r}) + 2 \times S \times 1s_A(\mathbf{r})1s_B(\mathbf{r})\right]$$

and I will leave you to prove that, for the simple H_2 LCAO wavefunction

$$\Psi_{\text{LCAO}}(\mathbf{r}_1,\mathbf{r}_2) = \frac{1}{2(1+S)}\left[(1s_A(\mathbf{r}_1) + 1s_B(\mathbf{r}_1))(1s_A(\mathbf{r}_2) + 1s_B(\mathbf{r}_2))\right]$$

and that the electron density is given by

$$P_1(\mathbf{r}) = \frac{1}{(1+S)}\left[1s_A^2(\mathbf{r}) + 1s_B^2(\mathbf{r}) + 2 \times 1s_A(\mathbf{r})1s_B(\mathbf{r})\right]$$

In either case, integration of $P_1(\mathbf{r})$ over the spatial variable gives 2, the number of electrons. In the general case of a many-electron wavefunction which will depend on the spatial variables $\mathbf{r}_1$, $\mathbf{r}_2$, ..., $\mathbf{r}_n$, we define

$$P_1(\mathbf{r}) = n\int \Psi^2(\mathbf{r}_1,\mathbf{r}_2\ldots\mathbf{r}_n)\,d\tau_2\ldots d\tau_n$$

where we have to calculate the integral and then write $\mathbf{r}$ in place of $\mathbf{r}_1$ in the answer. The electron density integrates to give the number of electrons, n.

Back to H_2 for a minute; when working with atomic orbitals, it is usual to rewrite the electron density in terms of a certain **matrix**, called (not surprisingly) the **electron density matrix**. What we do is collect together the atomic orbitals $1s_A(\mathbf{r})$ and $1s_B(\mathbf{r})$ into a 1×2 matrix $(1s_A(\mathbf{r})\,1s_B(\mathbf{r}))$. The electron density can then be written as a product of three matrices

$$P_1(\mathbf{r}_1) = (1s_A(\mathbf{r})\,1s_B(\mathbf{r}))\begin{pmatrix} \dfrac{1}{1+S^2} & \dfrac{S}{1+S^2} \\ \dfrac{S}{1+S^2} & \dfrac{1}{1+S^2} \end{pmatrix}\begin{pmatrix} 1s_A(\mathbf{r}) \\ 1s_B(\mathbf{r}) \end{pmatrix}$$

for the simple VB wavefunction and

$$P_1(\mathbf{r}) = (1s_A(\mathbf{r})\,1s_B(\mathbf{r})) \begin{pmatrix} \dfrac{1}{1+S} & \dfrac{1}{1+S} \\ \dfrac{1}{1+s} & \dfrac{1}{1+S} \end{pmatrix} \begin{pmatrix} 1s_A(\mathbf{r}) \\ 1s_B(\mathbf{r}) \end{pmatrix}$$

for the simple LCAO case. I am going to denote such matrices as $\underline{\mathbf{P}}_1$.

4.1 THE GENERAL LCAO CASE

Much time is spent these days in applications of the LCAO model to large molecules. To anticipate future chapters let's see how to calculate the $\underline{\mathbf{P}}_1$ matrix in the general case where we have a set of n atomic orbitals $\chi_1(\mathbf{r}), \chi_2(\mathbf{r}) \ldots \chi_n(\mathbf{r})$, and an LCAO molecular orbital

$$\psi_A(\mathbf{r}) = a_1\chi_1(\mathbf{r}) + a_2\chi_2(\mathbf{r}) + \ldots + a_n\chi_n(\mathbf{r}).$$

We can assume that the atomic orbitals are normalized to $1\left[\int \chi_1{}^2(\mathbf{r})d\tau = 1\right]$ but they need not be orthogonal.

The electron density for an electron of either spin in this spatial orbital is given by

$$P_1(\mathbf{r}) = (\chi_1(\mathbf{r})\,\chi_2(\mathbf{r}) \ldots \chi_n(\mathbf{r})) \begin{pmatrix} a_1{}^2 & a_1a_2 & \ldots & a_1a_2 \\ a_2a_1 & a_1{}^2 & \ldots & a_2a_n \\ \ldots & & & \\ a_na_1 & a_na_2 & \ldots & a_n{}^2 \end{pmatrix} \begin{pmatrix} \chi_1(\mathbf{r}) \\ \chi_2(\mathbf{r}) \\ \ldots \\ \chi_n(\mathbf{r}) \end{pmatrix}$$

and if the electronic state of that particular molecule could be written as $\psi_A^2\psi_B^2\psi_C^1$ we would calculate an electron density matrix having elements typically $2a_ia_j + 2b_ib_j + 1c_ic_j$ The 2s and the 1 are often called **occupation numbers**.

In view of our earlier discussion of the physical interpretation of the squares of wavefunctions, we know that $\int \psi_A^2(\mathbf{r})\,d\tau = 1$, and direct calculation of the integral shows that

$$1 = a_1{}^2 \int \chi_1^2(\mathbf{r})d\tau + a_1a_2 \int \chi_1(\mathbf{r})\chi_2(\mathbf{r})d\tau + \ldots + a_1a_n \int \chi_1(\mathbf{r})\chi_n(\mathbf{r})d\tau$$

$$+ a_2a_1 \int \chi_2(\mathbf{r})\chi_2(\mathbf{r})\,d\tau + a_2{}^2 \int \chi_2^2(\mathbf{r})\,d\tau + \ldots + a_2a_n \int \chi_2(\mathbf{r})\chi_n(\mathbf{r})\,d\tau$$

$$+ \ldots$$

$$+ a_na_1 \int \chi_n(\mathbf{r})\chi_1(\mathbf{r})\,d\tau + a_na_2 \int \chi_n(\mathbf{r})\chi_2(\mathbf{r})\,d\tau + \ldots + a_n{}^2 \int \chi_n^2(\mathbf{r})\,d\tau$$

The integrals over the atomic orbitals are usually collected together into an $n \times n$ matrix $\underline{\mathbf{S}}$ called the **overlap matrix**, and in a more compact notation

$$\sum\sum (\underline{\mathbf{P}}_1)_{ij}(\underline{\mathbf{S}})_{ji} = 1$$

In the more general case, where we have calculated the $\underline{\mathbf{P}}_1$ matrix according to the occupation numbers we have

$$\sum\sum(\underline{\mathbf{P}}_1)_{ij}(\underline{\mathbf{S}})_{ji} = \text{number of electrons}$$

4.2 POPULATION ANALYSIS

Once an approximation to the wavefunction of a molecule has been found, it can be used to calculate the probable result of many physical measurements, and hence to predict properties such as a molecular hexadecapole moment or the electric field gradient at a quadrupolar nucleus. For many workers in the field, this is the primary aim of performing quantum mechanical calculations. But from the early days of quantum chemistry, others have repeatedly tackled the problem of interpreting the wavefunction itself, attempting to understand why it takes a particular form for a particular Hamiltonian, and how the form of the wavefunction affects the expectation values calculated from it.

The importance of the density function $P_1(\mathbf{r})$ has been emphasized; it turns out that the expectation values of operators such as the electric dipole moment can be determined directly from P_1 without recourse to the wavefunction itself. The density function $P_1(\mathbf{r})$ has a simple quantum mechanical meaning; $P_1(\mathbf{r})\,d\tau$ gives the probability of finding an electron with either spin in the spatial volume element $d\tau$, and people often speak about $P_1(\mathbf{r})$ as if it were some kind of time average. Actually this interpretation has to be taken with caution; any single measurement on a quantum mechanical system changes the system. So we would need a very large number of systems prepared in identical electronic states before we could contemplate a time averaged experiment. The concept does not truly have a proper quantum mechanical interpretation.

For a preliminary survey of the electron density in a molecule, it is usual to make a pictorial representation of $P_1(\mathbf{r})$, as I showed you for $H_2{}^+$. Whilst such diagrams do not usually carry much information, they do provide a theoretical measure which can be compared to the results of experimental X-ray diffraction studies. A whole volume of the *Transactions of the American Chemical Society* was devoted to the Symposium 'Experimental and Theoretical Studies of Electron Densities', and Lipscomb's paper on 'Aspects of the electron density problem' is well worth reading (Lipscomb 1972).

But for the purposes of a purely theoretical analysis of molecular electronic structure, we need more detailed information. The term **population analysis** was introduced in a series of papers by Mulliken in 1955, but the basic ideas had already been anticipated by Mulliken himself and by other authors. The technique has been very widely applied since Mulliken's 1955 papers, because it is very simple and has the apparent virtue of being 'quantitative'. The word 'quantitative' seems to mean two different things to different authors:

Electronic Population Analysis on LCAO-MO Molecular Wave Functions I

R. S. Mulliken

Journal of Chemical Physics, **23** (1955) 1833–1840

With increasing availability of good all-electron LCAO-MO wavefunctions for molecules, a systematic procedure for obtaining maximum insight from such data has become desirable. An analysis in quantitative form is given here in terms of breakdowns of the electronic population into partial and total 'gross atomic populations' or into partial and total 'new atomic populations' together with 'overlap populations'. 'Gross atomic populations' distribute the electrons almost perfectly among the various AOs of the various atoms in the molecule. From these numbers, a definite figure is obtained for the amount of promotion (e.g. from 2s to 2p) in each atom; and also for the gross charge Q on each atom if the bonds are polar. The total overlap population for any pair of atoms in a molecule is in general made up of positive and negative contributions.

- an analytical description of the charge distribution in a molecule;
- a measure of the strength and nature of the bonding in a molecule.

Most users of population analysis seem to be concerned with the former. Take the LCAO treatment of H_2 as an example. We focus on the function

$$P_1(\mathbf{r}) = (1s_A(\mathbf{r})\,1s_B(\mathbf{r})) \begin{pmatrix} \dfrac{1}{1+S} & \dfrac{1}{1+S} \\ \dfrac{1}{1+S} & \dfrac{1}{1+S} \end{pmatrix} \begin{pmatrix} 1s_A(\mathbf{r}) \\ 1s_B(\mathbf{r}) \end{pmatrix}$$

and the matrix $\underline{\mathbf{P}}_1$ given by

$$\begin{pmatrix} \dfrac{1}{1+S} & \dfrac{1}{1+S} \\ \dfrac{1}{1+S} & \dfrac{1}{1+S} \end{pmatrix}$$

Atomic orbital $1s_A$ can obviously be associated with nucleus H_A and atomic orbital $1s_B$ with nucleus H_B. The first term in the electron density $1s_A{}^2(\mathbf{r})/(1+S)$ is taken to represent the amount of that electron density associated with nucleus H_A. The corresponding term $1s_B^2(\mathbf{r})/(1+S)$ represents the charge density associated with nucleus H_B. The remainder $2 \times 1s_B(\mathbf{r})1s_B(\mathbf{r})/(1+S)$ is taken to represent the electron density shared by the two nuclei. Mulliken's first idea was to integrate these contributions and so give the numbers $1/(1+S)$, $1/(1+S)$ and $2S/(1+S)$ in this specific case, and that these numbers would retain some chemical information.

The numbers are usually called the 'net atomic populations' and the 'overlap populations'. Chemists speak of the charges on atoms in molecules, and Mulliken's final contribution was to divide up the overlap regions between contributing atoms and allocate shares to the nuclei concerned. In the case of a homonuclear diatomic, there is no argument; the overlap region is divided equally between the two contributing atoms, giving **Mulliken gross atom populations** of $1/(1+S) + \frac{1}{2}(2S/(1+S)) = 1$ per atom. Remembering that these are electron densities, the overall charge on each atom is 0. Other authors have chosen different partitioning schemes.

4.3 DENSITY FUNCTIONS AND MATRICES

Let us now treat electron spin explicitly rather than integrating it out. We noted that for a one-electron wavefunction, $\Psi^2(\underline{r}, s)\,d\tau ds$ gives the chance of finding the electron in the spatial volume element $d\tau$ with spin between s and $s + ds$. Authors such as McWeeny and Sutcliffe (1969) write $\underline{x} = \underline{r}s$ to describe the space–spin variables, and I am going to follow this notation where it is appropriate. Both $\underline{r}$ and $\underline{x}$ are vectors in this notation.

When integrating, we normally have to integrate over space ($d\tau$) and spin (ds). (Some authors choose to write the differential element $d\tau ds$ as $d\underline{x}$, a vector quantity, when they really mean a scalar quantity). In the case of H_2^+, where there is only a single electron, $\Psi^2(\underline{x})$ is called the **density function** and it is written $\rho_1(\underline{x})$. The subscript '1' reminds us that we are dealing with the coordinates of a single electron.

For a many-electron system with wavefunction $\Psi(\underline{x}_1, \underline{x}_2 \ldots \underline{x}_n)$ then $\Psi^2(\underline{x}_1, \underline{x}_2 \ldots \underline{x}_n)\,d\tau_1 ds_1\, d\tau_2 ds_2 \ldots d\tau_n ds_n$ gives the probability of finding simultaneously electron 1 in $d\tau_1 ds_1$, electron 2 in $d\tau_2 ds_2 \ldots$ electron n in $d\tau_n ds_n$. The probability that electron 1 is in $d\tau_1 ds_1$ and the other electrons are anywhere is found by integrating this expression over the coordinates of electrons $2 \ldots n$ to give $d\tau_1 ds_1 \int \Psi^2(\underline{x}_1, \underline{x}_2 \ldots \underline{x}_n)\,d\tau_2 ds_2 \ldots d\tau_n ds_n$. Since the wavefunction product has to be symmetrical in the electron names, we define the one-electron density function as

$$\rho_1(\underline{x}_1) = n \int \Psi^2(\underline{x}_1, \underline{x}_2 \ldots \underline{x}_n)\,d\tau_2 ds_2 \ldots d\tau_n ds_n$$

where the $d\tau_1 ds_1$ refers to a point in spin-space where the density is to be evaluated. $\rho_1(\underline{x})$ is the first of a series of density functions which relate to clusters of any number of electrons. The next member of the series is

$$\rho_2(\underline{x}_1, \underline{x}_2) = n(n-1) \int \Psi^2(\underline{x}_1, \underline{x}_2 \ldots \underline{x}_n)\,d\tau_3 ds_3 \ldots d\tau_n ds_n$$

which determines the probability of any two electrons being found simultaneously at points $\underline{x}_1$ and $\underline{x}_2$. For an electronic wavefunction that is an eigenfunction of the spin operator S_z, it turns out that the one-electron density function always comprises an α^2 spin part and a β^2 spin part as follows

$$\rho_1(\underline{x}_1) = P_1^\alpha(\underline{r}_1)\alpha^2(s_1) + P_1^\beta(\underline{r}_1)\beta^2(s_1)$$

with no 'cross terms' involving $\alpha\beta$. The probability densities for plus spin electrons $\left[P_1^\alpha(\underline{r}_1) \right]$ and for minus spin electrons are always equal in a singlet spin state, but in non-singlet states the densities may be different, giving a resultant **spin density**. The spin density function is defined at points in space, and if we evaluate it at the position of certain nuclei, it gives a value proportional to the electron spin resonance isotropic hyperfine coupling constant.

4.4 DENSITY FUNCTIONAL THEORY

For electronic ground states, there exists a remarkable theory known as the **density functional theory.** This constitutes a method whereby we can work direct with $P_1(\underline{r})$ rather than with the many-electron wavefunction.

Quite a bit is known about the properties of the electron density $P_1(\underline{r})$; for example, the energy for a ground-state wavefunction is exactly determined by $P_1(\underline{r})$ alone. When the density is exactly equal to the true electron density, then the energy is a minimum. These two statements constitute the *Kohn–Hohenberg theorem* (Hohenberg and Kohn 1964). Many of the concepts of qualitative structural chemistry such as the Pauling electronegativity, Pearson's hardness and softness and Fukui's reactivity indices appear naturally from density functional theory.

There is a practical computation scheme called the Kohn–Sham equations (Kohn and Sham 1965). These are similar to the Hartree Fock equations, but they include electron correlation effects. There are several recent texts dealing with density functional theory, including those by March (1992) and by Parr and Yang (1989). Density functional theory may well prove to be more economical for problems where electron correlation is important than the traditional techniques, but this claim has yet to be demonstrated (1995). Commercially available packages now make a point of selling their 'DFT' options.

5 Self Consistent Fields

I discussed in Chapter 3 the concept of an idealized two-electron molecule H_2 where the electrons did not repel each other, and showed you how to deal with such a problem. The electronic Schrödinger equation separates into equations for either electron, and assuming that we can then solve the H_2^+ problem, H_2 holds no fears. The wavefunction for the two electrons is just a product of H_2^+ wavefunctions for either electron.

Once electron repulsion is taken into account, this separation of a many-electron wavefunction into a product of one-electron wavefunctions is no longer possible. This is not a failing of quantum modelling, scientists and engineers reach the same conclusion whenever they have to deal with problems involving three or more mutually interacting particles. We talk about the 'three-body problem'.

Whilst astronomers also suffer from the three body problem, they are lucky in that gravitational interactions depend on the product of **masses** rather than the product of **charges**, and so models of the solar system are based on the massive attractions between our sun and the planets. The gravitational interactions between the planets themselves can then be treated as small perturbations on the dominant gravitational interaction between the sun and the planets.

The 'one electron model' (i.e. the orbital model) is a very attractive one, and obviously can be used to model atoms and molecules since it is now part of the language of elementary descriptive chemistry. How do we go about recovering this 'single-electron' model when dealing with a many-electron system?

The essence of the **Hartree Fock (HF)** or **self consistent field (SCF)** model is to solve the electronic Schrödinger equation for a single electron moving in a potential where the complicated motions of the remaining electrons are 'averaged out'. These interactions are certainly not taken to be zero, but the model obviously lacks the finer details caused by the instantaneous electron–electron interactions. The basic physical idea is therefore a simple one and can be tied in nicely with our discussion of the electron density of Chapter 4. We noted the physical significance of the density function $\rho_1(\underline{x})$; $\rho_1(\underline{x})d\tau ds$ gives the chance of finding any electron in the spin–space volume element $d\tau ds$ with the other electrons anywhere and with either spin. $P_1(\mathbf{r})d\tau$ gives the corresponding chance of finding any electron with either spin in the spatial volume $d\tau$.

There are several ways in which we can proceed. The traditional one is to look for an eigenvalue equation for the HF orbitals. Another way is to concentrate on the electron density itself and to seek methods for finding $P_1(\mathbf{r})$ direct, without recourse to the wavefunction.

The vast majority of known molecules are organic ones, totally lacking in symmetry. The vast majority of these have singlet electronic ground states, which can be written as molecular orbital configurations $\psi_A^2 \psi_B^2 \ldots \psi_M^2$ and visualized as shown in Figure 5.1.

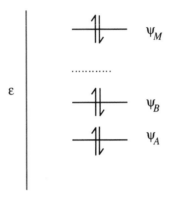

Figure 5.1   Orbital level diagram for a typical organic molecule

We speak about M occupied molecular orbitals, and the number of electrons is $2M$, because we have allocated an α and a β spin electron to each molecular orbital. In the original Hartree model, the many-electron wavefunction was written as a straightforward product of one-electron functions

$$\Psi_{el}(\underline{\mathbf{x}}_1, \underline{\mathbf{x}}_2, \ldots \underline{\mathbf{x}}_{2M}) = \psi_A(\mathbf{r}_1)\alpha(s_1) \quad \psi_A(\mathbf{r}_2)\beta(s_2) \ldots \psi_M(\mathbf{r}_{2M})\beta(s_{2M})$$

and I will refer to the Hartree model from time to time in the text.

With the hindsight of 'indistinguishability', you should know that the smallest possible logical building block for such a system is

$$\Psi_{el}(\underline{\mathbf{x}}_1, \underline{\mathbf{x}}_2, \ldots \underline{\mathbf{x}}_{2M}) = \begin{vmatrix} \psi_A(\mathbf{r}_1)\alpha(s_1) & \psi_A(\mathbf{r}_1)\beta(s_1)\ldots & \psi_M(\mathbf{r}_1)\beta(s_1) \\ & & \ldots \\ \psi_A(\mathbf{r}_{2M})\alpha(s_{2M}) & \psi_A(\mathbf{r}_{2M})\beta(s_{2M})\ldots & \psi_M(\mathbf{r}_{2M})\beta(s_{2M}) \end{vmatrix}$$

where we have doubly occupied each SCF orbital and taken account of antisymmetry by writing a Slater determinant. This is known as the Hartree Fock (HF) model. From now on, I will use the terms SCF and HF interchangeably.

I am assuming that this particular electronic state is the lowest energy one of the given symmetry, and that the ψ's are orthonormal. The first assumption is a vital one, the second just makes the algebra a little easier. I haven't normalized the total wavefunction Ψ_{el}, but will need to remember this fact later in the derivation. Traditionally we now minimize the variational energy

$$\epsilon_{el} = \frac{\int \Psi_{el} H_{el} \Psi_{el} d\tau}{\int \Psi_{el}^2 d\tau}$$

The integration is over the coordinates of all the electrons, and all the ψ's are assumed to be real quantities rather than complex ones.

The aim of HF theory is to find the best 'form' of the electronic wavefunction, which is taken as a product of one-electron functions. Each electron moves in an average field due to the nuclei and the remaining electrons. The first step is to work out the formal energy expression in terms of the one and two-electron operators and the HF orbitals ψ_R. For a polyatomic, polyelectronic molecule with $2M$ electrons and NUC nuclei, the electronic Hamiltonian is a sum of terms representing

- the kinetic energy of each of the $2M$ electrons (which are denoted by subscript i)
- the mutual potential energy of each nucleus (NUC in total, subscript α) with every electron
- the mutual potential energy of each pair of electrons

and in symbols this is

$$H_{el} = -\frac{h^2}{8\pi^2 m} \sum_{i=1}^{2M} \nabla_i^2 - \frac{e^2}{4\pi\epsilon_0} \sum_{i=1}^{2M} \left(\sum_{\alpha=1}^{NUC} \frac{Z_\alpha}{R_{\alpha,i}} \right)$$
$$+ \frac{e^2}{4\pi\epsilon_0} \sum_{i=1}^{2M-1} \left(\sum_{j=i+1}^{2M} \frac{1}{r_{ij}} \right)$$

I have grouped the operators together for a reason. We normally simplify the notation along the lines already discussed for H_2, and write the electronic Hamiltonian as a sum of the one-electron operators and two-electron operators

$$H_{el} = \sum_{i=1}^{2M} h(\mathbf{r}_i) + \sum_{i=1}^{2M-1} \left[\sum_{j=i+1}^{2M} g(\mathbf{r}_i, \mathbf{r}_j) \right]$$

already discussed. They are defined as

$$h(\mathbf{r}_i) = -\frac{h^2}{8\pi^2 m} \nabla_i^2 - \frac{e^2}{4\pi\epsilon_0} \left(\sum_{\alpha=1}^{NUC} \frac{Z_\alpha}{R_{\alpha,i}} \right)$$
$$g(\mathbf{r}_i, \mathbf{r}_j) = \frac{e^2}{4\pi\epsilon_0} \frac{1}{r_{ij}}$$

The energy expression can now be found using the Slater Condon Shortley rules. These rules are given in all the classic texts such as EWK. The idea is that the energy expression which involves the coordinates of **all** electrons can be reduced to a much simpler sum of terms involving the coordinates of one and (at most) two electrons. The electronic energy can be written in terms of the ψ's and the one and two-electron operators as

$$
\begin{aligned}
\epsilon_{\text{el}} = {} & 2 \sum_{R=A}^{M} \int \psi_R(\mathbf{r}_1) h(\mathbf{r}_1) \psi_R(\mathbf{r}_1) \mathrm{d}\tau_1 \\
& + \sum_{R=A}^{M} \sum_{S=A}^{M} (2 \int \int \psi_R(\mathbf{r}_1) \psi_R(\mathbf{r}_1) g(\mathbf{r}_1, \mathbf{r}_2) \psi_S(\mathbf{r}_2) \psi_S(\mathbf{r}_2) \mathrm{d}\tau_1 \mathrm{d}\tau_2 \\
& - \int \int \psi_R(\mathbf{r}_1) \psi_S(\mathbf{r}_1) g(\mathbf{r}_1, \mathbf{r}_1) \psi_R(\mathbf{r}_2) \psi_S(\mathbf{r}_2) \mathrm{d}\tau_1 \mathrm{d}\tau_2)
\end{aligned}
$$

The first of the two electron terms has a classical physical significance. To see this, you need to remember that $-e\psi_R^2(\mathbf{r}_1)$ is the charge density associated with the electron we have labelled 1 and $-e\psi_S^2(\mathbf{r}_2)$ is the charge density for electron 2. The integral is therefore the mutual electrostatic potential energy of the pair of electrons. The second two-electron term is a little more difficult to understand; formally, it represents the mutual electrostatic potential energy of the 'overlap' charge distribution $-e\psi_R(\mathbf{r}_1)\psi_S(\mathbf{r}_1)$ (due to electron 1) with an identical term $-e\psi_R(\mathbf{r}_2)\psi_S(\mathbf{r}_2)$ due to electron 2.

We now attempt to find the minimum of this quantity by varying the form of the ψ values subject to the constraint that they remain orthonormal. I will explain this in detail shortly, but I must mention that atoms are very special beasts because of their symmetry.

All the early work was concerned with atoms, and Sir William Hartree is regarded as the founder of SCF theory (Hartree 1927). His son, D. R. Hartree published the definitive book, *The Calculation of Atomic Structures* in 1957, in which he derived the atomic Hartree Fock (HF) equations and described numerical procedures for their solution (Hartree 1957). As I will explain shortly, the HF orbitals ψ_i are found as solutions of the HF eigenvalue problem, such that

$$
h^{\text{F}} \psi_i = \epsilon_i \psi_i
$$

Charlotte Froese Fischer was supervised in her research by D. R. Hartree, and she published **her** definitive book, *The Hartree Fock Method for Atoms: A Numerical Approach* in 1977. The appendix lists a number of freely available atomic structure programs. Most of these can be obtained from the Computer Physics Communications Program Library.

For atoms, solution of the HF equations to full accuracy is routine. We refer to such calculations as being at the 'Hartree Fock Limit'. This represents the best solution possible within the independent electron model. For large molecules, solutions at the HF limit are not possible.

5.1 THE LCAO PROCEDURE

In fact, the HF procedure leads to a complicated set of near intractable integro-differential equations, which can only be solved with any ease for a one-centre problem. If you are interested in the atomic case, you should read the classic books mentioned above. What we normally do for molecules is to use the LCAO model where each molecular orbital can be described as a linear combination of n atomic orbitals $\chi_1, \chi_2, \ldots, \chi_n$. This formulation of the HF equations, where we use the LCAO approximation, is usually credited to Roothaan (1951a). Each electron with position vector $\underline{r}$ is described by an LCAO orbital written as a linear combination of the χ's, e.g.

$$\psi_A(\underline{r}) = a_1\chi(\underline{r}) + a_2\chi(\underline{r}) + \ldots a_n\chi_n(\underline{r})$$

and I now want to derive an expression for the variational energy in terms of the LCAO coefficients, the one and two-electron operators and the χ's.

A neat derivation involves the density matrix $\underline{P}_1(\underline{r})$ of Chapter 4. I showed you how to calculate the charge density $P_1(\underline{r})$ and the matrix $\underline{P}_1$ from the LCAO coefficients in the case of a singly occupied MO $\psi_{A;}$.

$$P_1(\underline{r}) = (\chi_1(\underline{r})\chi_2(\underline{r}) \ldots \chi_n(\underline{r})) \begin{pmatrix} a_1^2 & a_1 a_2 & \ldots & a_1 a_n \\ a_2 a_1 & a_1^2 & \ldots & a_2 a_n \\ \ldots & & & \\ a_n a_1 & a_n a_2 & \ldots & a_n^2 \end{pmatrix} \begin{pmatrix} \chi_1(\underline{r}) \\ \chi_2(\underline{r}) \\ \ldots \\ \chi_n(\underline{r}) \end{pmatrix}$$

where the square matrix in this case is $\underline{P}_1$.

In the case of M doubly occupied LCAO ψ values, the elements of the matrix $\underline{P}_1$ are given by

$$\underline{P}_{ij} = 2 \sum_{r=a}^{r=m} r_i r_j$$

An easy way to remember this (provided that you are *au fait* with matrices) is to collect the LCAO coefficients for all the M doubly occupied orbitals into an $n \times M$ matrix $\underline{U}$

$$\underline{U} = \begin{pmatrix} a_1 & b_1 & \ldots & m_1 \\ a_2 & b_2 & \ldots & m_2 \\ \ldots & & & \\ a_n & b_n & \ldots & m_n \end{pmatrix}$$

and then $\underline{P}_1 = 2\underline{U}\underline{U}^T$ where $\underline{U}^T$ is the transpose of $\underline{U}$.

$$\underline{U}^T = \begin{pmatrix} a_1 & a_2 & \ldots & a_n \\ b_1 & b_2 & \ldots & b_n \\ \ldots & & & \\ m_1 & m_n & \ldots & m_n \end{pmatrix}$$

(Just to remind you how to calculate the product of two matrices; if $\underline{\mathbf{A}}$ has n rows and l columns, and $\underline{\mathbf{B}}$ has l rows and p columns, then $\underline{\mathbf{AB}}$ has n rows and p columns and the ij element is

$$(\underline{\mathbf{AB}})_{ij} = \sum_{k=1}^{l} A_{ik} B_{kj}$$

Don't get the LCAO orbitals (the ψ's) get mixed up with the atomic orbitals (the χ's). Most authors refer to the χ values as **basis functions**, and I will follow the practice.

The atomic orbitals usually overlap each other, and I am going to collect together the basis function overlap integrals $\int \chi_i(\underline{\mathbf{r}})\chi_j(\underline{\mathbf{r}})d\underline{\mathbf{r}}$ into an $n \times n$ matrix $\underline{\mathbf{S}}$.

$$\begin{pmatrix} S_{1,1} & S_{1,2} & \cdots & S_{1,n} \\ S_{2,1} & S_{2,2} & \cdots & S_{2,n} \\ \cdots & & & \\ S_{n,1} & S_{n,2} & \cdots & S_{n,n} \end{pmatrix}$$

We rarely work with orthogonal basis functions χ. The LCAO orbitals ψ are **usually** orthogonal and this fact can be summarized in a simple matrix statement. A little manipulation will show you that $\underline{\mathbf{U}}^{\mathrm{T}}\underline{\mathbf{S}}\underline{\mathbf{U}}$ is a unit matrix (with m rows and m columns), and also that

$$\underline{\mathbf{P}}_1\underline{\mathbf{S}}\underline{\mathbf{P}}_1 = 4\underline{\mathbf{P}}_1$$

So, we now use the variation principle to seek the best possible values of the LCAO coefficients. Remember that for a molecule we are working within the Born–Oppenheimer approximation and so the nuclei are clamped at their positions in space. What I have to do is to calculate the electronic variational integral ϵ_{el} as above and then 'optimize' it (by setting its first derivative to zero). The optimization thus involves calculating the first order change in energy with respect to all the variables, and then setting this first order change to zero and solving the resultant equation(s). I could work with the LCAO coefficients, but a neater way is to work directly with the charge density matrix $\underline{\mathbf{P}}_1$. We keep track of the requirement that the LCAO ψ's have to be orthonormal by requiring that

$$\underline{\mathbf{P}}_1\underline{\mathbf{S}}\underline{\mathbf{P}}_1 = 4\underline{\mathbf{P}}_1.$$

If I collect together all integrals involving the atomic orbitals $\chi_1(\underline{\mathbf{r}}), \chi_2(\underline{\mathbf{r}}), \ldots, \chi_n(\underline{\mathbf{r}})$ and a typical one-electron operator into an $n \times n$ matrix $\underline{\mathbf{h}}_1$ such that

$$(\underline{\mathbf{h}}_1)_{ij} = \int \chi_1(\underline{\mathbf{r}})h(\underline{\mathbf{r}})\chi_1(\underline{\mathbf{r}})d\tau$$

then the one-electron operators turn out to make a contribution

$$\sum_{i=1}^{n}\sum_{j=1}^{n}(\underline{\mathbf{h}}_1)_{ij}(\underline{\mathbf{P}}_1)_{ji}$$

to the variational energy. This double sum can be written a little more neatly as the **trace** of the matrix product $\underline{\mathbf{P}}_1\underline{\mathbf{h}}_1$ (or $\underline{\mathbf{h}}_1\underline{\mathbf{P}}_1$). [The **trace** of a square $n \times n$ matrix $(A)_{ij}$ is the sum of its diagonal elements, the product $\underline{\mathbf{P}}_1\underline{\mathbf{h}}_1$ is $n \times n$ and the diagonal elements of the product are $\sum (h_1)_{ij}(P_1)_{ji}$ with the sum over j.] The two-electron contribution to the energy can be written as $\frac{1}{2}\mathrm{trace}(\underline{\mathbf{P}}_1\underline{G})$, where the elements of the matrix $\underline{G}$ depends on the elements of the matrix $\underline{\mathbf{P}}_1$ in a complicated way;

$$G_{ij} = \sum\sum P_{1,kl}\Big(\int\int \chi_i(\underline{\mathbf{r}}_1)\chi_j(\underline{\mathbf{r}}_1)g(\underline{\mathbf{r}}_1,\underline{\mathbf{r}}_2)\chi_k(\underline{\mathbf{r}}_2)\chi_l(\underline{\mathbf{r}}_2)d\tau_1 d\tau_2$$
$$-\frac{1}{2}\int\int \chi_i(\underline{\mathbf{r}}_1)\chi_k(\underline{\mathbf{r}}_1)g(\underline{\mathbf{r}}_1,\underline{\mathbf{r}}_2)\chi_j(\underline{\mathbf{r}}_2)\chi_l(\underline{\mathbf{r}}_2)d\tau_1 d\tau_2\Big)$$

The total electronic energy is finally given by

$$\epsilon_{\mathrm{el}} = \mathrm{trace}\,(\underline{\mathbf{h}}_1\underline{\mathbf{P}}_1) + \frac{1}{2}\mathrm{trace}\,(\underline{\mathbf{P}}_1\underline{G})$$

The next step in the variation calculation is to examine how the energy changes when the electron density changes. We put $\underline{\mathbf{P}}_1 \to \underline{\mathbf{P}}_1 + \delta\underline{\mathbf{P}}_1$ and after a little manipulation find that the first order change in the energy is

$$\delta\epsilon_{\mathrm{el}} = \mathrm{trace}\,(\underline{\mathbf{h}}_1\delta\underline{\mathbf{P}}_1) + \mathrm{trace}(\delta\underline{\mathbf{P}}_1\underline{G})$$

Note that the factor of $\frac{1}{2}$ has disappeared from the energy expression; this is because the $\underline{G}$ matrix itself depends on the electron density, and this has to be taken into account.

We write the $\delta\epsilon_{\mathrm{el}}$ equation in terms of the Hartree Fock Hamiltonian matrix

$$\underline{\mathbf{h}}^{\mathrm{F}} = \underline{\mathbf{h}}_1 + \underline{G}$$

so that $\delta\epsilon_{\mathrm{el}} = \mathrm{trace}\,(\delta\underline{\mathbf{P}}_1\underline{\mathbf{h}}^{\mathrm{F}})$. For the record, and because we will refer to them again and again, the elements of the HF matrix are

$$h_{ij}^{\mathrm{F}} = \int \chi_i(\underline{\mathbf{r}}_1)h(\underline{\mathbf{r}}_1)\chi_j(\underline{\mathbf{r}}_1)d\tau_1$$
$$+ \sum\sum P_{1,kl}[\int\int \chi_i(\underline{\mathbf{r}}_1)\chi_j(\underline{\mathbf{r}}_1)g(\underline{\mathbf{r}}_1,\underline{\mathbf{r}}_2)\chi_k(\underline{\mathbf{r}}_2)\chi_l(\underline{\mathbf{r}}_2)d\tau_1 d\tau_2]$$
$$-\frac{1}{2}\int\int \chi_i(\underline{\mathbf{r}}_1)\chi_k(\underline{\mathbf{r}}_1)g(\underline{\mathbf{r}}_1,\underline{\mathbf{r}}_2)\chi_j(\underline{\mathbf{r}}_2)\chi_l(\underline{\mathbf{r}}_2)d\tau_1 d\tau_2)$$

We want to find a change $\delta\underline{\mathbf{P}}_1$ that makes the energy variation $\delta\epsilon_{\mathrm{el}}$ equal to zero. $\underline{\mathbf{P}}_1$ contains information about the number of electrons, and it also satisfies the condition that

$$\underline{P}_1 \underline{S} \underline{P}_1 = 4\underline{P}_1$$

(which is the same thing as saying that the LCAO orbitals are orthonormal). We substitute for the new $\underline{P}_1$ into this equation and a little manipulation will show you that, at the energy minimum

$$\underline{h}^F \underline{P}_1 = \underline{P}_1 \underline{h}^F$$

This shows that, when we have found the correct electron density matrix $\underline{P}_1$ and calculated the correct Hartree Fock Hamiltonian $\underline{h}^F$ from it, the two matrices will satisfy the condition that $\underline{h}^F \underline{P}_1 = \underline{P}_1 \underline{h}^F$ (we say that they **commute**).

This doesn't help us immediately to **find** the electron density matrix (remember that $\underline{h}^F$ is defined in terms of $\underline{P}_1$), but it tells us the condition for a minimum in the variational energy. What we often do is this:

- have a 'guess' at the electron density $\underline{P}_1$
- calculate $\underline{h}^F$ from that $\underline{P}_1$
- check on the condition for a minimum
- improve $\underline{P}_1$ somehow, until this condition is met.

A lot of effort has gone into devising procedures for the final step, and you might like to read about the steepest descents procedure in McWeeny and Sutcliffe (1969).

But you are probably wondering about the LCAO orbitals and their LCAO coefficients? An alternative formulation of the LCAO self consistent field problem in terms of these coefficients is that of Roothaan. It works in terms of the eigenvalues and eigenvectors of the Hartree Fock hamiltonian matrix. In terms of a typical LCAO orbital describing an electron with position $\mathbf{r}$

$$\psi_A(\mathbf{r}) = a_1 \chi(\mathbf{r}) + a_2 \chi(\mathbf{r}) + \ldots a_n \chi n(\mathbf{r})$$

then at the energy minimum the coefficients a_i form an eigenvector of $\underline{h}^F$, and the lowest energy eigenvectors correspond to the occupied LCAO-MOs. So if we write the LCAO coefficients as a column vector

$$\underline{a}^T = (a_1, a_2 \ldots a_n)$$

$$\underline{h}^F \underline{a} = \epsilon_A \underline{S} \underline{a}$$

where ϵ_A is called the **orbital energy**.

Roothaan's procedure thus involves finding the matrix $\underline{h}^F$, which still depends on the LCAO-MO coefficients. The procedure is pretty much the same:

- Have a 'guess' at the LCAO-MO coefficients
- Calculate $\underline{h}^F$ from these LCAO-MO coefficients

- Check on the condition for a minimum
- Improve the LCAO-MO coefficients somehow, until this condition is met.

Again, a great deal of effort has gone into devising methods to solve the final step. We needn't worry about the details, and I will show you the SCF procedure in operation in a later chapter.

5.1.1 The electronic energy

Suppose then that we have a molecule with $2M$ electrons, constrained to occupy M LCAO-MOs, and that the SCF orbitals have energies $\epsilon_A, \epsilon_B \ldots \epsilon_M$. The LCAO coefficients are also collected in the column vectors $\mathbf{a}, \mathbf{b} \ldots \mathbf{m}$. The electronic energy is given by

$$\epsilon_{el} = \text{trace}\,(\underline{\mathbf{h}}_1 \underline{\mathbf{P}}_1) + \frac{1}{2}\text{trace}\,(\underline{\mathbf{P}}_1 \mathbf{G})$$

and the sum of the orbital energies is given by

$$\epsilon_{orb} = 2(\epsilon_A + \epsilon_B + \ldots + \epsilon_M)$$

Now

$$\underline{\mathbf{h}}^F \underline{\mathbf{a}} = \epsilon_A \underline{\mathbf{S}}\underline{\mathbf{a}}$$

and so

$$\underline{\mathbf{a}}^T \underline{\mathbf{h}}^F \underline{\mathbf{a}} = \epsilon_A \underline{\mathbf{a}}^T \underline{\mathbf{S}}\underline{\mathbf{a}} = \epsilon_A$$

and with a little manipulation you will find that

$$\epsilon_{el} = \epsilon_{orb} - \frac{1}{2}\text{trace}(\underline{\mathbf{P}}_1 \mathbf{G})$$

so that the orbital energies do not sum to the total electronic energy.

5.2 KOOPMANS' THEOREM

The orbital energies for such a doubly occupied SCF wavefunction don't have any simple significance in terms of the electronic energy, but they do have significance as follows. Suppose we remove a single electron from orbital ψ_X, and estimate the energy required for this process. Koopmans found that, provided the ψ's do not change on ionization, the ionization energy is related simply to the SCF orbital energy ϵ_X by

$$\text{ionization energy} = -\epsilon_X$$

This result was derived (Koopmans 1934) for an exact SCF wavefunction in the numerical Hartree Fock sense, but it is also valid for a wavefunction calculated using the LCAO approximation. Unfortunately, the molecular orbitals describing the resulting cation might well be quite different from those of the parent molecule because the electron density might relax on ionization. So we had better examine how to calculate the wavefunctions for simple excited states within the HF model.

5.3 OPEN SHELLS

The vast majority of molecules are electronic singlet states, and they can be described adequately by the above procedure. There **are** electronic states, particularly for simple molecules of high symmetry, where the single Slater determinant approach is not valid. I gave you the example of the H_2 singlet state earlier.

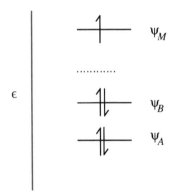

Figure 5.2   An ROHF (Restricted Open Shell Hartree Fock) situation

The self consistent field approach is very attractive since it gives an orbital picture, and it turns out that several classes of open shell electronic states can still be treated. The simplest case is that shown, where we have a number of doubly occupied orbitals and a number of orbitals that are singly occupied but with all their spins parallel. We often speak of the 'closed shell' and the 'open shell', and in Figure 5.2 I have shown a case with only one open shell electron.

We could treat the triplet state of H_2 using this procedure, but not the singlet state. The algebraic treatment is very similar to the closed shell case, except that we now need to consider an electron density matrix for the closed shell electrons and an electron density matrix for the open shell electrons, and at the end of the calculation we get a set of LCAO coefficients and orbital energies. I'll give a concrete example in a little while.

5.4 UNRESTRICTED HARTREE FOCK THEORY

A more general way to treat systems having an odd number of electrons (many atoms and most radicals and ions) is to let the individual LCAOs become **singly** occupied, as in Figure 5.3.

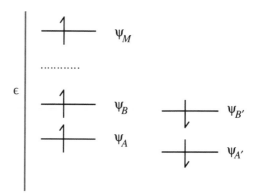

Figure 5.3   An unrestricted Hartree Fock situation

In standard SCF theory, we constrain the wavefunction so that every LCAO is doubly occupied. The idea of **unrestricted Hartree Fock theory** (UHF) is to allow the α and the β spin electrons to have different spatial wavefunctions. To determine the LCAO coefficients, we can minimize the energy as before. There are two points to notice:

- the α and the β electrons are considered separately and there is an electron density matrix for each set
- the resulting wavefunction is not necessarily an eigenfunction of the spin operator S^2 and this may be very important depending on the particular application.

6 Hückel Theory

Molecules with extensive π bonding systems, like benzene, are not described well by classical valence bond theory. Such molecules are often planar, or largely planar, and two types of bond contribute to their structure, σ bonds and π bonds. The σ bonds are to be thought of as **localized** between atoms, whilst the π bonds are to be thought of as being **delocalized** over large portions of the molecule. Much of the interesting chemistry of such compounds appears to relate to the so-called π electrons, and we think about these π electrons moving in a potential caused by the nuclear framework and the σ electrons. From the early days of quantum chemistry, people have striven to give simple models of such compounds in an effort to

- explain the special structure, stability and properties of these molecules
- to predict new properties of new molecules.

One of the earliest models for π-electron molecules is afforded by **Hückel π-electron theory**, which dates from the 1930s (Hückel 1931). The ideas are simple and appealing, and the model enjoyed many years of successful application to individual molecules, clusters and solids. Imagine a simple π-electron molecule such as ethene, and let's concentrate as usual on the electron motion. There are 16 electrons in total ($14\sigma + 2\pi$), so our total electronic wavefunction will be some complicated function of their **spatial** and **spin** variables $\underline{x}_1, \underline{x}_2, \ldots, \underline{x}_{16}$. We write this mathematically as $\Psi(\underline{x}_1, \underline{x}_2, \ldots, \underline{x}_{16})$. Don't confuse this **state wavefunction** $\Psi(\underline{x}_1, \underline{x}_2, \ldots, \underline{x}_{16})$ with a **molecular orbital**! We might well build up the state wavefunction $\Psi(\underline{x}_1, \underline{x}_2, \ldots, \underline{x}_{16})$, which describes all 16 electrons, from molecular orbitals each of which describes a single electron, and to do this we would probably use a Slater determinant.

Where would these one-electron molecular orbitals come from? I explained in Chapter 5 the basic ideas of the **Hartree Fock self consistent field method**; we find each of these molecular orbitals by solving the Hartree Fock Hamiltonian problem

$$h^{\mathrm{F}} \psi_i = \epsilon_i \psi_i$$

You should remember the basic physical idea behind the HF method; each electron moves in an **average** field due to the other electrons (and of course the nuclei), so that the HF operator contains within itself the averaged electron density due to the other electrons. I also mentioned the LCAO version of Hartree Fock theory where we seek to expand individual molecular orbitals ψ_i in terms of some fixed 'basis functions'; if we denote an LCAO-MO by ψ, and the atomic orbital basis function set by $\chi_1, \chi_2, \ldots, \chi_n$ then we write

$$\psi = c_1\chi_1 + c_2\chi_2 + \ldots + c_n\chi_n$$

and we once again normally use the variation technique to find the LCAO-MO coefficients $c_1, c_2, \ldots, c_n$. Each LCAO-MO, of course, describes a **single** electron, and we have to build up a state wavefunction from the LCAO-MOs in terms of Slater determinants.

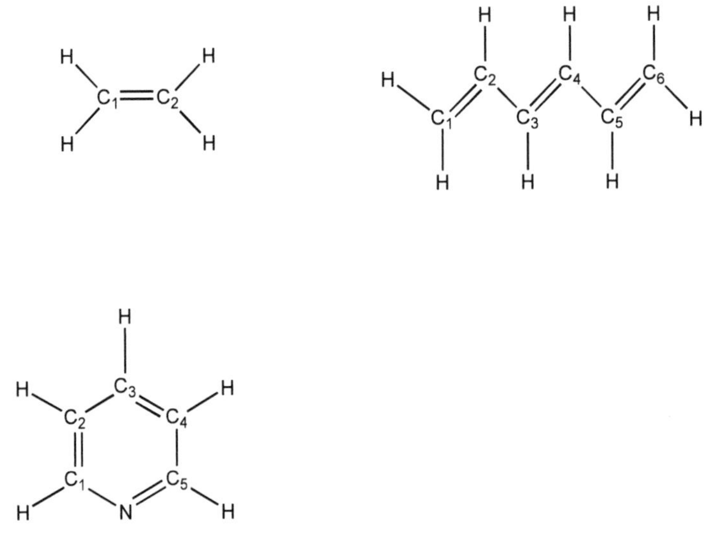

Figure 6.1   The three molecules discussed: ethene, hexatriene, pyridine

To find the LCAO-MO coefficients, we have to solve the matrix LCAO-HF equations

$$\underline{h}^F \underline{a} = \epsilon_A \underline{S}\underline{a}$$

where $\underline{h}^F$ is now the Hartree Fock Hamiltonian matrix, and $\underline{S}$ is the matrix of basis function overlap integrals. Written out in full,

$$\begin{pmatrix} h^{\mathrm{F}}_{1,1} & h^{\mathrm{F}}_{1,2} & \ldots & h^{\mathrm{F}}_{1,n} \\ h^{\mathrm{F}}_{2,1} & h^{\mathrm{F}}_{2,2} & \ldots & h^{\mathrm{F}}_{2,n} \\ \ldots & & & \\ h^{\mathrm{F}}_{n,1} & h^{\mathrm{F}}_{n,2} & \ldots & h^{\mathrm{F}}_{n,n} \end{pmatrix} \begin{pmatrix} a_1 \\ a_2 \\ \ldots \\ a_n \end{pmatrix} = \epsilon_A \begin{pmatrix} S_{1,1} & S_{1,2} & \ldots & S_{1,n} \\ S_{2,1} & S_{2,2} & \ldots & S_{2,n} \\ \ldots & & & \\ S_{n,1} & S_{n,2} & \ldots & S_{n,n} \end{pmatrix} \begin{pmatrix} a_1 \\ a_2 \\ \ldots \\ a_n \end{pmatrix}$$

For the minute, imagine an LCAO-SCF treatment of ethene which treats only the π electrons. Each first-row atom in the conjugated system is assumed to contribute one atomic orbital of the correct symmetry to the conjugated system, and without any particular justification other than chemical intuition we make the following approximations.

> **The π atomic orbitals are orthogonal and normalized**

so that each LCAO-MO consists of a linear combination of those atomic orbitals having the correct symmetry (e.g. carbon $2p_\pi$ AOs). Not only that, the matrix elements in the Hartree Fock Hamiltonian matrix equations are assumed to be constants.

> **'Atom' parameters α_X depend only on atom type X**
> **'Bond' parameters β_{XY} depend only on atoms X and Y**
> **Non-bonded interactions ignored**

We write these α_C and β_{CC} for carbon atoms and carbon–carbon conjugated bonds respectively. Interactions between non-bonded atoms are set to zero. The physical interpretation is that each of the π electrons moves in some kind of average potential due to the σ-electron core, the nuclei and the remaining π electron(s).

6.1 EXAMPLES

6.1.1 Ethene

Ethane is now a two-electron problem; each π electron moves in an average potential due to the remaining π electron, the σ electrons and the nuclear framework. Thus we want to find LCAO-MOs to describe the two π electrons in terms of the two carbon $2p_\pi$ atomic orbitals on the two constituent carbon atoms making up the framework. See Figure 6.1 for atom numbering. Note that I have numbered the carbon atoms and that we take single atomic $2p_\pi$ orbitals on each carbon centre. If we write these atomic orbitals χ_1, χ_2 and make the Hückel approximations, the LCAO-HF equations become

$$\begin{pmatrix} \alpha & \beta \\ \beta & \alpha \end{pmatrix} \begin{pmatrix} c_1 \\ c_2 \end{pmatrix} = \epsilon \begin{pmatrix} c_1 \\ c_2 \end{pmatrix}$$

No iterations are necessary in order to solve these SCF equations, because the left hand matrix only involves constants. Nevertheless, the equations are probably different from ones you have been used to dealing with.

The problem is referred to by mathematicians as the **algebraic eigenvalue problem**. I don't want to bore you with details, and have given you a simple BASIC program called 'JACOBI' on your disk. If you don't have the disk, you will be easily able to get hold of a matrix diagonalization program suitable for the problem. Such programs expect as input a matrix of order n, whose elements are **numbers** not **symbols** such as α and β. We could always substitute experimental values for α and β, but there is no need. If we divide the secular equations by β and put $x = (\alpha - \epsilon)/\beta$, then the matrix to be input is just,

$$\begin{pmatrix} 0 & 1 \\ 1 & 0 \end{pmatrix}$$

Output consists of the eigenvalues and eigenvectors of this matrix. In LCAO language these are the orbital energies (in terms of x rather than ϵ) and the LCAO coefficients. I have arranged the LCAO MOs in order of increasing energy (it turns out that β is a negative quantity).

Table 6.1 Hückel π-electron treatment of ethene

Orbital energy ϵ	c_1	c_2
$\alpha + \beta$	0.7071	0.7071
$\alpha - \beta$	0.7071	−0.7071

6.1.2 Hexatriene

Hexatriene (Figure 6.1) is now a 6π-electron problem; each π electron moves in an average potential due to the remaining five π electrons, the σ electrons and the nuclear framework. Thus we want to find LCAO-MOs to describe the six π electrons in terms of the six carbon $2p_\pi$ atomic orbitals on the atoms making up the framework. If we write these atomic orbitals $\chi_1, \chi_2, \ldots, \chi_6$ then the LCAO- SCF equations become

$$\begin{pmatrix} \alpha & \beta & 0 & 0 & 0 & 0 \\ \beta & \alpha & \beta & 0 & 0 & 0 \\ 0 & \beta & \alpha & \beta & 0 & 0 \\ 0 & 0 & \beta & \alpha & \beta & 0 \\ 0 & 0 & 0 & \beta & \alpha & \beta \\ 0 & 0 & 0 & 0 & \beta & \alpha \end{pmatrix} \begin{pmatrix} c_1 \\ c_2 \\ c_3 \\ c_4 \\ c_5 \\ c_6 \end{pmatrix} = \epsilon \begin{pmatrix} c_1 \\ c_2 \\ c_3 \\ c_4 \\ c_5 \\ c_6 \end{pmatrix}$$

Note the zeros in the left-hand matrix; if carbon atoms i and j are not bonded, then the matrix element is zero and so we put a zero in the $1, 3$ the $1, 4$ (etc.) positions.

Now do a Hückel calculation for hexatriene (or examine the second example on the Hückel program on your disk). You will find the orbital energies and LCAO coefficients given in Table 6.2. Notice the molecular symmetry.

Table 6.2 Hückel π-electron treatment of hexatriene

x	c_1	c_2	c_3	c_4	c_5	c_6
1.802	0.2319	0.4179	0.5211	0.5211	0.4179	0.2319
1.247	−0.4179	−0.5211	−0.2319	0.2319	0.5211	0.4179
0.445	0.5211	0.2319	−0.4179	−0.4179	0.2319	0.5211
−0.445	−0.5211	0.2319	0.4179	−0.4179	−0.2319	0.5211
−1.247	−0.4179	0.5211	−0.2319	−0.2319	0.5211	−0.4179
−1.802	−0.2319	0.4179	−0.5211	0.5211	−0.4179	0.2319

6.1.3 Bond lengths and the Hückel method

In the electronic ground state of the molecule, the six π electrons occupy the three lowest energy orbitals (the first three in Table 6.2). The π-electron density matrix $\underline{\mathbf{P}}_1$ turns out to be

$$\underline{\mathbf{P}}_1 = \begin{pmatrix} 1.0000 & 0.8711 & 0.0000 & -0.3877 & 0.0000 & 0.3014 \\ 0.8711 & 1.0000 & 0.4834 & 0.0000 & -0.863 & 0.0000 \\ 0.0000 & 0.4834 & 1.0000 & 0.7849 & 0.0000 & -0.3877 \\ -0.3877 & 0.000 & 0.7849 & 1.0000 & 0.4834 & 0.0000 \\ 0.0000 & -0.0863 & 0.0000 & 0.4834 & 1.0000 & 0.8711 \\ 0.3014 & 0.0000 & -0.3877 & 0.0000 & 0.8711 & 1.0000 \end{pmatrix}$$

Notice that the shorter bonds have a higher overlap density (the 'bond order'). It was noted a long time ago that there was a linear relationship between bond length R and π-electron bond order p, usually written

$$R/\text{pm} = 150 - 16p$$

This is discussed in detail in Streitwieser's book (Streitwieser 1962). More sophisticated relations have also been proposed.

6.1.4 Molecular mechanics of π-electron systems

Many MM packages now give special consideration to π-electron systems for the following reason. MM force fields are usually parameterized according to the 'classical valence' description of a given heavy atom: C(sp), C(sp^2) and C(sp^3), depending on the local symmetry of the atom. It was found that this gave a poor representation to the enthalpies of formation of conjugated systems and so the following correction is often made. We make a Hückel calculation on the input geometry and then adjust the bond lengths in accordance with the calculated bond orders. The MM calculation is then made on this adjusted geometry in a self consistent fashion.

6.1.5 Treatment of heteroatoms

How does the Hückel Molecular Orbital (HMO) method deal with pyridine? We need to account for the fact that N is more electronegative than C, and so the N atom should have a higher π-electron (negative) charge than an average C atom. It is conventional to write heteroatom parameters such as α_N in terms of the 'known' parameters α and β (sometimes written as α_C and β_{CC}) thus

$$\alpha_N = \alpha_C + h_N \beta_{CC}$$

and simple arguments based on the π-electron 'polarizability' show that h_N should be positive. Different authors recommend different values ranging from 0.5 to 1.2. Obviously, the N atom in pyrrole must have a different value than the N atom in pyridine. Some suggested values are given in Table 6.3.

Table 6.3

Atom X	$h(X)$	$k(C-X)$
C	1.0	1.0
N (donates 1 π)	0.5	0.8
N (donates 2 π)	1.5	1.0
B	-1.0	0.7
O (donates 1 π)	1.0	0.8
Methyl	2.0	0.7

Similar arguments have been advanced to cope with C-C and C-X bonds of differing types. We write (for example)

$$\beta_{CN} = k_{CN} \beta_{CC}$$

where the parameter k has to be found by calibrating theoretical versus experimental results. Typical values are also given in the Table 6.3.

6.1.6 Pyridine

The secular equations for pyridine (Figure 6.1) are

$$\begin{pmatrix} \alpha_N & \beta_{CN} & 0 & 0 & 0 & \beta_{CN} \\ \beta_{CN} & \alpha & \beta & 0 & 0 & 0 \\ 0 & \beta & \alpha & \beta & 0 & 0 \\ 0 & 0 & \beta & \alpha & \beta & 0 \\ 0 & 0 & 0 & \beta & \alpha & \beta \\ \beta_{CN} & 0 & 0 & 0 & \beta & \alpha \end{pmatrix} \begin{pmatrix} c_1 \\ c_2 \\ c_3 \\ c_4 \\ c_5 \\ c_6 \end{pmatrix} = \epsilon \begin{pmatrix} c_1 \\ c_2 \\ c_3 \\ c_4 \\ c_5 \\ c_6 \end{pmatrix}$$

and with a choice of $\beta_{CN} = \beta_{CC}$ and $\alpha_N = \alpha_C + \beta_{CC}$ calculate the HMOs (or examine the third example on your disk).

Contrary to popular belief, Hückel π-electron theory is not dead and buried. Papers appear from time to time dealing with topics such as dielectric susceptibilties (McIntyre and Hameka 1978) and soliton dynamics (Su and Schrieffer 1980). The latest version of MOBY, v1.5, has just **added** HMO theory as an option.

6.2 EXTENDED HÜCKEL THEORY

Hückel calculations on planar systems were extensively exploited, and I refer you to Streitwieser's classic book *Molecular Orbital Theory for Organic Chemists*, (Streitwieser 1961). What few calculations had been done on the σ framework had used the method of linear combination of bond orbitals.

The Spectra and Electronic Structure of the Tetrahedral Ions
MnO_4^-, CrO_4^{2-} and ClO_4^-
Max Wolfsberg and Lindsay Helmholtz
Journal of Chemical Physics, **20** (1952) 837–843

We have made use of a semiempirical treatment to calculate the energies of the molecular orbitals for the ground state and the first few excited states of permanganate, chromate and perchlorate ions. The calculation of the excitation energies is in agreement with the qualitative features of the observed spectra, i.e. absorption in the far ultraviolet for ClO_4^-, two strong maxima in the visible or near ultraviolet for MnO_4^- and CrO_4^{2-} with the chromate spectrum displaced towards higher energies. An approximate calculation of the relative f-values for the first transitions in CrO_4^{2-} and MnO_4^- is also in agreement with experiment.

The data on the absorption spectra of permanganate ion in different crystalline fields is interpreted in terms of the symmetries of the excited states predicted by our calculations.

The pioneering calculations of Wolfsberg and Helmholtz are usually cited as the first applications of 'extended' Hückel theory.

The extended Hückel method treats valence electrons within the spirit of the π electron variant. Each LCAO-MO is written as a linear combination of valence orbitals $\chi_1, \chi_2, \ldots \chi_n$ and once again we have to solve the SCF equations

$$\underline{h}^F \underline{a} = \epsilon_A \underline{S} \underline{a}$$

where $\underline{h}^F$ is now a valence-electron Hartree Fock Hamiltonian matrix, and $\underline{S}$ is the matrix of atomic orbital overlap integrals.

$$\begin{pmatrix} h_{1,1}^{\text{F}} & h_{1,2}^{\text{F}} & \dots & h_{1,n}^{\text{F}} \\ h_{2,1}^{\text{F}} & h_{2,2}^{\text{F}} & \dots & h_{2,n}^{\text{F}} \\ \dots & & & \\ h_{n,1}^{\text{F}} & h_{n,2}^{\text{F}} & \dots & h_{n,n}^{\text{F}} \end{pmatrix} \begin{pmatrix} a_1 \\ a_2 \\ \dots \\ a_n \end{pmatrix} = \epsilon \begin{pmatrix} S_{1,1} & S_{1,2} & \dots & S_{1,n} \\ S_{2,1} & S_{2,2} & \dots & S_{2,n} \\ \dots & & & \\ S_{n,1} & S_{n,2} & \dots & S_{n,n} \end{pmatrix} \begin{pmatrix} a_1 \\ a_2 \\ \dots \\ a_n \end{pmatrix}$$

Within the spirit of Hückel theory, the 'core' of inner shell electrons and the nuclei are assumed to be averaged into the elements of the HF matrix, and these elements are chosen according to Mulliken's recipe:

$$h_{i,j}^{\text{F}} = K(h_{i,i}^{\text{F}} + h_{j,j}^{\text{F}})S_{i,j}/2$$

The critical choices are the parameters K and the diagonal elements of the HF matrix. According to Mulliken, the latter can be given the physical interpretation that they represent the energy of an electron in the ith atomic orbital, in the field of the nuclear skeleton and remaining electrons.

Wolfsberg and Helmholtz chose values for Cr not greatly different from the valence state ionization energies for Cr^0, and estimated the remainder.

The systematic application of extended Hückel theory to organic molecules comes with the work of Hoffmann:

An Extended Hückel Theory. I. Hydrocarbons
Roald Hoffmann
The Journal of Chemical Physics, **39** (1963) 1397–1412

The Hückel theory, with an extended basis set consisting of 2s and 2p carbon and 1s hydrogen orbitals, with inclusion of overlap and all interactions, yields a good qualitative solution of most hydrocarbon conformational problems. Calculations have been performed within the same parameterization for nearly all simple saturated and unsaturated compounds, testing a variety of geometries for each. Barriers to internal rotation, ring conformations, and geometrical isomerism are among the topics treated. Consistent σ and π charge distributions and overlap populations are obtained for aromatics and their relative roles discussed. For alkanes and alkenes charge distributions are also presented. Failures include overemphasis on steric factors, which leads to some incorrect isomerization energies; also the failure to predict strain energies. It is stressed that the geometry of a molecule appears to be its most predictable quality.

Many molecular modelling packages now include extended Hückel theory as an option. You might be interested to learn how overlap integrals are calculated. The classic work is

Formulas and Numerical Tables for Overlap Integrals
R. S. Mulliken, C. A. Rieke, D. Orloff and H. Orloff
Journal of Chemical Physics, **17** (1949) 1248–1267

Explicit formulas and numerical tables for the overlap integral S between AOs (atomic orbitals) of two overlapping atoms a and b are given. These cover all the most important combination of AO pairs involving ns, $np\sigma$ and $np\pi$ AOs. They are based on approximate AOs of the Slater type, each containing two parameters μ [equal to $Z/(n-\delta)$ and $n-\delta$ where $n-\delta$ is an effective principal quantum number. The S formulas are given as functions of two parameters p and t, where $p = \frac{1}{2}(\mu_a + \mu_b)R/a_0$, R being the internuclear distance, and $t = (\mu_a + \mu_b)/(\mu_a - \mu_b)$. Master tables of computed values of S are given over wide ranges of p and t values corresponding to actual molecules, and also including the case $p = 0$ (intra-atomic overlap integrals) . . .

6.3 'THE NIGHTMARE OF THE INNER SHELLS'

At the beginning of this Chapter, I introduced the idea that the 16 electrons in ethene could be divided conceptually into two sets, the 14σ and the 2π electrons. I referred to the electrons individually by the variables $\underline{x}_1, \underline{x}_2, \ldots \underline{x}_{16}$, and for the minute I am going to formally labelled the π electrons 1 and 2, and the σ electrons 3 through 16 (although they are, of course, indistinguishable, and any acceptable wavefunction has to take account of this fact). Formally, methods such as the Hückel π-electron model aim to solve the electronic Schrödinger equation for the π electrons moving in some effective field due to the nuclei and the remaining σ-electron core.

The first step in the problem is to rewrite the electronic Hamiltonian as a 'core' part, a π-electron part and the remainder, which will represent the interaction between the π electrons and the 'core', which includes the nuclei.

The electronic Hamiltonian for the 16 electrons is easily written down

$$H_{\text{el}} = -\frac{h^2}{8\pi^2 m}\sum_{i=1}^{16}\nabla_i^2 - \frac{e^2}{4\pi\epsilon_0}\sum_{i=1}^{16}\left(\sum_{\alpha=1}^{6}\frac{Z_\alpha}{R_{\alpha,i}}\right) + \frac{e^2}{4\pi\epsilon_0}\sum_{i=1}^{15}\left(\sum_{j=i+1}^{16}\frac{1}{r_{ij}}\right)$$

and it can be separated formally into a 'core part' (electrons 3 through 16), a 'π-electron part' (electrons 1 and 2), and the remaining two-electron terms which describe the interaction between the two groups of electrons.

$$H_{\text{el}} = H_{\text{core}} + H_{\pi} + H_{\text{interaction}}$$

where, for example, $$H_{\text{core}} = -\frac{h^2}{8\pi^2 m}\sum_{i=3}^{16}\nabla_i^2$$

$$-\frac{e^2}{4\pi\epsilon_0}\sum_{i=3}^{16}\left(\sum_{\alpha=1}^{6}\frac{Z_\alpha}{R_{\alpha,i}}\right)$$

$$+\frac{e^2}{4\pi\epsilon_0}\sum_{i=3}^{15}\left(\sum_{j=i+1}^{16}\frac{1}{r_{ij}}\right)$$

If it weren't for the interaction terms, then we could go ahead and concentrate on each set of electrons independently.

Hückel π-electron theory gives a simple way of calculating the π wavefunction, $\Psi_\pi(\underline{x}_1, \underline{x}_2)$. Imagine now that the σ-electron wavefunction has been calculated somehow and that it is $\Psi_\sigma(\underline{x}_3, \underline{x}_4, \ldots \underline{x}_{16})$. Each of the two wavefunctions will satisfy the generalized Pauli principle, namely, that they must change sign on interchanging the names of two electrons. We might be tempted to write a product wavefunction

$$\Psi_\sigma(\underline{x}_3, \underline{x}_4, \ldots \underline{x}_{16})\Psi_\pi(\underline{x}_1, \underline{x}_2)$$

to describe the total, but we would then run into a problem; the total wavefunction has to change sign if we interchange the names of two π electrons (which it does), of two σ electrons (which it does), and any σ electron with any π electron (which at the minute it doesn't). We have to allow for this in constructing the full wavefunction. Without going into details, we do this by partially antisymmetrizing the product; we write the full wavefunction (Antisymmetrizing operator) $\Psi_\sigma(\underline{x}_3, \underline{x}_4, \ldots \underline{x}_{16})\Psi_\pi(\underline{x}_1, \underline{x}_2)$

Finally, a full analysis shows that it is indeed possible to treat **either** group of electrons as if they were moving in some kind of average field due to the nuclei and the other group. To do this, we introduce certain 'effective' one-electron operators $h^{\text{eff}}(\mathbf{r})$, which can actually be rigorously defined. We therefore make variational calculations on (e.g.) the two π electrons in ethene with an effective electronic Hamiltonian

$$H_{\text{el}}^{\text{eff}} = \sum_{i=13}^{14}(h^{\text{eff}}\mathbf{r}_i) + \frac{e^2}{4\pi\epsilon_0}\left(\frac{1}{r_{1,2}}\right)$$

Obviously, Hückel theory goes way beyond this and approximates the entire matrix elements. But during the early days of quantum mechanics, it soon became customary to simplify the problem of molecular binding between many-electron atoms by treating the valence electrons alone, regarding the atomic cores as simply localized positive charges. This valence electron approximation gave remarkably satisfactory results and, at first, there seemed to be no justifiable mathematical grounds for the approximation. Van Vleck and Sherman (1935) referred to the problem as the 'nightmare of the inner shells'.

6.4 BUT WHAT *IS* THE HÜCKEL HAMILTONIAN?

A basic tenet of Hückel π-electron theory is that the orbital energies add to give the π electron energy. But the Hartree Fock orbital energies do **not** add in this way, as I showed you earlier. So strictly, the Hückel Hamiltonian is not the Hartree Fock Hamiltonian. The interpretation usually accepted is that the Hückel Hamiltonian matrix should be identified with the matrix $\underline{\mathbf{h}}^F - \frac{1}{2}\underline{\mathbf{G}}$, where $\underline{\mathbf{G}}$ is the repulsion matrix. The basis for this belief is that the matrix $\underline{\mathbf{h}}^F - \frac{1}{2}\underline{\mathbf{G}}$ has eigenvalues that **do** sum to give the correct electronic energy. This is discussed in detail, in McWeeny and Sutcliffe (1969).

7 Differential Overlap Models

Hückel models of molecular electronic structure enjoyed many years of popularity, particularly the π-electron variants. Perhaps, because nothing else was technically feasible at the time, authors sought to extract the last possible pieces of information from these models. Thus for example, the inductive effect was described. The inductive effect is a cornerstone concept in organic chemistry; a group R should show a $+I$ or a $-I$ effect (according to the nature of the group R) when it is added to a benzene ring.

Wheland and Pauling (1959) tried to explain the inductive effect in terms of π-electron theory by varying the atomic α and the bond β parameters for nearest neighbour atoms, next nearest neighbour atoms . . . But as many authors have also pointed out, it is always easy to introduce yet more parameters into a simple model, obtain agreement with some experimental finding and then claim that the model represents some kind of absolute truth.

Of course you will get excellent agreement with experimental results if you add more and more parameters to a model. But what about the transferability of these parameters from molecule to molecule? Or even from property to property within the same molecule? This is the stuff of chemistry. Not only that, but electron repulsion is not treated explicitly; the simple Hückel models average somehow over electron repulsion. Our Hückel treatment of pyridine was discussed in an earlier Chapter 6. The electronic configuration of pyridine in its ground state is $\psi_1^2 \psi_2^2 \psi_3^2$ and the lowest excited configurations in the Hückel model would be written with orbital configuration $\psi_1^2 \psi_2^2 \psi_3^1 \psi_4^1$, irrespective of the resultant electron spin.

The Hückel orbital energies from Chapter 6 are $\alpha + 2.278\beta$, $\alpha + 1.317\beta$, $\alpha + 1.000\beta$, $\alpha - 0.705\beta$, $\alpha - 1.000\beta$ and $\alpha - 1.891\beta$. This means that the first excited $\pi \rightarrow \pi^*$ triplet and singlet states would each have an energy $(\alpha - 0.705\beta) - (\alpha + 1.000\beta) = -1.705\beta$ above the electronic ground state (remember that β is a negative quantity). Experimentally, triplet excitation energies are usually significantly lower than singlet ones because of the effect of electron repulsion. So this represents a big failure of the simple π-electron models.

A second failure is obviously the treatment of $n \to \pi^*$ transitions, and I will return to this later in the Chapter.

7.1 THE π-ELECTRON MODELS

The next step came in the 1950s, with a more serious attempt to include formally the effect of electron repulsion within valence electron models. First came the π-electron models associated with the names of Pople, and with Pariser and Parr. You might like to read the synopses for their first papers.

Electron Interaction in Unsaturated Hydrocarbons
J. A. Pople
Transactions of the Faraday Society, **49** (1953) 1375–1385

An approximate form of the molecular orbital theory of unsaturated hydrocarbon molecules in their ground states is developed. The molecular orbital equations rigorously derived from the correct many-electron Hamiltonian are simplified by a series of systematic approximations and reduce to equations comparable with those used in the semi-empirical method based on an incompletely defined one-electron Hamiltonian. The two sets of equations differ, however, in that those of this paper include certain important terms representing electronic interaction. The theory is used to discuss the resonance energies, ionization potentials, charge densities, bond orders and bond lengths of some simple hydrocarbons. The electron interaction terms introduced in the theory are shown to play an important part in determining the ionization potentials. It is also shown that the uniform charge density theorem, proved by Coulson and Rushbrooke for the simpler theory, holds also for the self-consistent orbitals derived by the method of this paper.

A Semi-Empirical Theory of the Electronic Spectra and Electronic Structure of Complex Unsaturated Molecules
I. R. Pariser and R. G. Parr
Journal of Chemical Physics, **21** (1953) 466–471

A semi-empirical theory is outlined which is designed for the correlation and prediction of the wavelengths and intensities of the first main visible or ultraviolet bands and other properties of complex unsaturated molecules, and preliminary application of the theory is made to ethylene and to benzene.

   The theory is formulated in the language of the purely theoretical method of antisymmetrized products of molecular orbitals (in LCAO approximation) including configuration interaction, but departs from this theory in several essential respects. First, atomic orbital integrals involving the core Hamiltonian are expressed in terms of quantities which may be regarded as semi-empirical. Second, an approximation of zero differential overlap is employed, and an optionally uniformly charged sphere representation of atomic π-orbitals is introduced, which greatly simplify the evaluation of electronic repulsion integrals and make applications to complex molecules containing heteroatoms relatively simple. Finally, although the theory starts from the π-electron approximation, in which the unsaturation electrons are treated apart from the rest, provision is included for the adjustment of the σ-electrons to the π-electron distribution in a way which does not complicate the mathematics . . .

In honour of these authors, we speak of the PPP (Pariser–Parr–Pople) method, and to see what is involved, let's discuss our pyridine treatment in detail. Once again, we appeal to chemical intuition and divide the 42 electrons into two groups, the π electrons (6 in total) and the σ electron core (36 in total). The spirit of the PPP models is to treat the effect of the σ electrons as some kind of empirical constants whose values have to be determined by appeal to experiment. At the time, this was the wiser course! (and it probably still is).

As you will have noted, Pople's original paper was couched in terms of Hartree Fock language, so let's go down that path and examine the specific approximations used. Remember that we are using pyridine as the example. Just as in Hückel π-electron theory, the atomic orbitals used are not rigorously defined, but can be imagined as carbon $2p_\pi$ orbitals (for the moment). There are six π electrons, and we consider the ground electronic state with orbital configuration $\psi_A^2\psi_B^2\psi_C^2$. We use a single $2p_\pi$-type atomic orbital per conjugated atom, and so a PPP SCF calculation on pyridine therefore involves solving the (6×6) matrix SCF problem

$$\underline{\mathbf{h}}^{\mathrm{F}}\underline{\mathbf{a}} = \epsilon_A \underline{\mathbf{S}}\underline{\mathbf{a}}$$

It is usual in PPP theory to assume that the atomic orbitals involved (the χ's) are both normalized and orthogonal, so that the overlap matrix $\underline{\mathbf{S}}$ is the unit matrix. Obviously, ordinary atomic $2p_\pi$ atomic orbitals are **not** orthogonal, but just live with this dilemma for now.

7.1.1 Zero differential overlap

As I told you earlier, a typical element of the SCF matrix involves both one-electron and two-electron terms. The two electron terms are by far the most difficult, so let's start there. The i,j element of the G matrix is

$$G_{ij} = \sum\sum P_{1,kl}\left(\int\int \chi_i(\mathbf{r}_1)\chi_j(\mathbf{r}_1)g(\mathbf{r}_1,\mathbf{r}_2)\chi_k(\mathbf{r}_2)\chi_l(\mathbf{r}_2)\mathrm{d}\tau_1\mathrm{d}\tau_2\right.$$
$$\left. -\frac{1}{2}\int\int \chi_i(\mathbf{r}_1)\chi_k(\mathbf{r}_1)g(\mathbf{r}_1,\mathbf{r}_2)\chi_j(\mathbf{r}_2)\chi_l(\mathbf{r}_2)\mathrm{d}\tau_1\mathrm{d}\tau_2\right)$$

and PPP theory makes the **zero differential overlap** approximation; whenever a product of atomic orbitals appears under an integral sign in a two-electron integral

$$\int \cdots\cdots\cdots \chi_i(\mathbf{r}_1)\chi_j(\mathbf{r}_1)d\tau_1$$

then the integral is taken as **zero** if $i \neq j$, and a **constant** when $i = j$ depending only on the atom that the atomic orbital is centred on. Thus, a typical repulsion integral

$$\int\int \chi_i(\mathbf{r}_1)\chi_j(\mathbf{r}_1)g(\mathbf{r}_1,\mathbf{r}_2)\chi_k(\mathbf{r}_2)\chi_l(\mathbf{r}_2)\, d\tau_1 d\tau_2$$

is zero unless $i = j$ and $k = l$. These non-zero integrals are traditionally given the symbol γ.

We could, in principle, calculate the value of these integrals, if we knew exactly which atomic orbitals were involved. When Pariser and Parr first tried to calculate the excitation energies of unsaturated hydrocarbons on the assumption that the basis functions χ_i were $2p_\pi$ Slater atomic orbitals, they got very poor agreement with experimental results (see the landmark paper). But when they treated the integrals as parameters that could be fixed by appeal to experiment, they got much better agreement.

Thus, within the spirit of PPP theory, Zero Differential Overlap (ZDO) repulsion integrals are treated as parameters to be ultimately fixed by reference to experimental results. They are written γ_{ij}, and they are taken to depend on the nature of the two orbitals involved and the internuclear separation.

Over the years, a number of different parameter schemes were tried and the simplest one is probably that due to Mataga and Nishimoto, who wrote

$$\gamma_{ij} = \frac{e^2}{4\pi\epsilon_0}\frac{1}{R_{ij} + a_{ij}}$$

This is the expression you would get for the mutual electrostatic potential energy of a pair of charges each of magnitude $(-e)$, separated by a distance $R_{ij} + a_{ij}$ (the a term is essentially a parameter).

Now for the one-electron terms. We want to calculate integrals such as

$$\int \chi_i(\mathbf{r}_1)h(\mathbf{r}_1)\chi_j(\mathbf{r}_1)d\tau_1 = -\frac{h^2}{8\pi^2 m}\int \chi_i(\mathbf{r}_1)\nabla_1^2\chi_j(\mathbf{r}_1)d\tau_1$$
$$-\frac{e^2}{4\pi\epsilon_0}\int \chi_i(\mathbf{r}_1)\sum_{\alpha=A}^{NUC}\frac{Z_\alpha}{R_{\alpha,1}}\chi_j(\mathbf{r}_1)d\tau_1$$

Remember that orbital χ_i is centred on nucleus i, which contributes Z_i electrons to the conjugated system ($Z_i = 2$ for the oxygen atom in furan C_4H_4O).

Let me treat 'diagonal' terms, where $i = j$, and 'off-diagonal' terms (where $i \neq j$) separately. Off-diagonal terms are taken to depend on the type and length of the bond joining atoms i and $\neq j$, but not on the neighbouring bonds or atoms. The entire integral is taken to be a constant β_{ij}, which is **not** the same as the β in Hückel theory. This β_{ij} is usually treated as a basic parameter in π-electron ZDO theories, and it is usually set to zero for atoms that are not formally bonded. The diagonal terms need a little more consideration; they depend on the atom i and its neighbours.

Diagonal terms

$$\int \chi_i(\mathbf{r}_1)h(\mathbf{r}_1)\chi_i(\mathbf{r}_1)\mathrm{d}\tau_1 = -\frac{h^2}{8\pi^2 m}\int \chi_i(\mathbf{r}_1)\nabla_1^2\chi_i(\mathbf{r}_1)\mathrm{d}\tau_1$$
$$-\frac{e^2}{4\pi\epsilon_0}\int \chi_i(\mathbf{r}_1)\sum_{\alpha=A}^{NUC}\frac{Z_\alpha}{R_{\alpha,1}}\chi_i(\mathbf{r}_1)\mathrm{d}\tau_1$$

are expanded to separate out a contribution from the nucleus on which atomic orbital χ_i is centred, and the remaining nuclei

$$\int \chi_i(\mathbf{r}_1)h(\mathbf{r}_1)\chi_i(\mathbf{r}_1)\mathrm{d}\tau_1 = \left[-\frac{h^2}{8\pi^2 m}\int \chi_i(\mathbf{r}_1)\nabla_1^2\chi_i(\mathbf{r}_1)\mathrm{d}\tau_1 - \frac{e^2}{4\pi\epsilon_0}\int \chi_i(\mathbf{r}_1)\frac{Z_i}{R_{A,1}}\chi_i(\mathbf{r}_1)\mathrm{d}\tau_1\right]$$
$$-\frac{e^2}{4\pi\epsilon_0}\int \chi_i(\mathbf{r}_1)\sum_{\beta\neq A}\frac{Z_\beta}{R_{\beta,1}}\chi_i(\mathbf{r}_1)\mathrm{d}\tau_1$$

The first term in brackets on the right hand side is usually taken as the **valence state ionization energy** written ω_i, which is invariably treated as a parameter to be determined by appeal to experimental results, and the remaining terms are given values $-Z_j\gamma_{ij}$.

To summarize then, a ZDO π-electron SCF calculation on a molecule such as pyridine would therefore involve calculating the Hartree Fock matrix with elements

$$h_{ij}^F = \omega_i + \frac{1}{2}P_{1,ii}\gamma_{ii} + \sum_{j\neq i}(P_{1,jj} - Z_j)\gamma_{ij}$$
$$h_{ij}^F = \beta_{ij} - \frac{+1}{2}P_{1,ij}\gamma_{ij}$$

and repeating the iterations until the charge density matrix was constant. I have enclosed a simple demonstration BASIC program as an Appendix; the same three examples as for the Hückel theory are given. I have used the Mataga and Nishimoto parameter scheme, for the sake of illustration.

In order to perform a PPP model calculation on pyridine, we need the geometry of the π-electron framework. I won't enter into any debate as to what geometry should be used in these calculations but, traditionally, people chose to make life simple and would normally take idealized structures that were as regular as possible. So conjugated CC bond lengths are usually taken as 1.4 Å and all ring structures are treated as regular. Output from this program consists of orbital energies, the LCAO-MO coefficients and charge density data geometry.

The pyridine LCAO-MO output is given in Table 7.1; ϵ_i is the SCF energy of orbital i, and the c values are the LCAO coefficients. Atom number 1 is nitrogen.

Table 7.1   PPP π-electron treatment of pyridine

ϵ/eV	c_1	c_2	c_3	c_4	c_5	c_6
−13.988	−0.562	−0.424	−0.337	−0.313	−0.337	−0.424
−11.127	0.539	0.165	−0.371	−0.617	−0.371	0.165
−10.394	0	−0.495	−0.505	0	0.505	0.495
−1.275	0.518	−0.349	−0.267	0.588	−0.267	−0.349
−0.907	0	0.505	−0.495	0	0.495	−0.505
+1.965	0.353	−0.415	0.422	−0.420	0.422	−0.415

You probably noticed the theme of the prediction of electronic spectra in the landmark papers. This is a very difficult problem, and workers at the time (including myself) tackled it at the time with a certain naivety. Here is how we did it.

In SCF theory, the ground electronic configuration of pyridine is written $\psi_A^2\psi_B^2\psi_C^2$, which we would have to describe by a single Slater determinant. I will write this SCF wavefunction as Ψ_0. The simplest way we can describe the electronic excited states is with similar wavefunctions formed where an electron has been promoted from the occupied SCF orbital (e.g.) ψ_A to the SCF virtual orbital ψ_X, which I will write $\Psi_{(A \to X)}$. These are the so-called singly excited states and we would have to write them as a combination of (two) Slater determinants in order to describe correctly the singlet excited spectroscopic states. There are nine such possible states in this model, which arise by promoting a single electron from each of the three filled SCF orbitals to each of the three virtual ones, and together with Ψ_0 they give our first approximation to the ground and excited spectroscopic singlet states of pyridine.

The relative energies of these wavefunctions can be calculated within the PPP approximations and are shown in Table 7.2.

Table 7.2   Relative pyridine energies within the PPP approximation

State	Energy/eV
Ψ_0	0
$\Psi(3 \to 4)$	5.697
$\Psi(2 \to 5)$	6.544
$\Psi(3 \to 5)$	6.591
$\Psi(2 \to 4)$	6.897
$\Psi(3 \to 6)$	8.689
$\Psi(1 \to 4)$	9.095
$\Psi(1 \to 5)$	9.253
$\Psi(2 \to 6)$	9.327
$\Psi(1 \to 6)$	11.977

The next step would be to perform a configuration interaction calculation with these state wavefunctions.

Brillouin's theorem (Brillouin 1933) tells us that the singly excited states do not interact with the SCF singly excited states

$$\int \Psi_0 H \Psi(A \rightarrow X) d\tau = 0$$

The Hamiltonian matrix elements

$$\int \Psi(A \rightarrow X) H \Psi(B \rightarrow Y) d\tau$$

can be evaluated in terms of the integrals over atomic orbitals and the LCAO coefficients, using the Slater–Condon–Shortley rules. You might like to learn how this is done.

In the case where the LCAO orbitals have been determined by the SCF procedure, these matrix elements can be written elegantly in terms of the SCF orbital energies as follows. The diagonal elements of this matrix are

$$\int \Psi(A \rightarrow X) H \Psi(A \rightarrow X) d\tau = \epsilon_X - \epsilon_A$$

$$- \int \int \psi_A(\mathbf{r}_1) \psi_A(\mathbf{r}_1) g(\mathbf{r}_1, \mathbf{r}_2) \psi_X(\mathbf{r}_2) \psi_X(\mathbf{r}_2) d\tau_1 d\tau_2$$

$$+ 2 \int \int \psi_A(\mathbf{r}_1) \psi_X(\mathbf{r}_1) g(\mathbf{r}_1, \mathbf{r}_2) \psi_A(\mathbf{r}_2) \psi_X(\mathbf{r}_2) d\tau_1 d\tau_2$$

and the off-diagonal elements are

$$\int \Psi(A \rightarrow X) H \Psi(A \rightarrow X) d\tau = - \int \int \psi_A(\mathbf{r}_1) \psi_B(\mathbf{r}_1) g(\mathbf{r}_1, \mathbf{r}_2) \psi_X(\mathbf{r}_2) \psi_Y(\mathbf{r}_2) d\tau_1 d\tau_2$$

$$+ 2 \int \int \psi_A(\mathbf{r}_1) \psi_X(\mathbf{r}_1) g(\mathbf{r}_1, \mathbf{r}_2) \psi_B(\mathbf{r}_2) \psi_Y(\mathbf{r}_2) d\tau_1 d\tau_2$$

One of the time-consuming parts of a CI calculation is the transformation of the electron repulsion integrals calculated over the basis functions to integrals calculated over the LCAO-MOs. So for example, a typical two-electron integral

$$\int \int \psi_A(\mathbf{r}_1) \psi_B(\mathbf{r}_1) g(\mathbf{r}_1, \mathbf{r}_2) \psi_X(\mathbf{r}_2) \psi_Y(\mathbf{r}_2) d\tau_1 d\tau_2$$

would have to be evaluated by somehow performing the sum over basis functions χ according to

$$\sum \sum \sum \sum a_i b_j x_k y_l \int \int \chi_i(\mathbf{r}_1) \chi_j(\mathbf{r}_1) g(\mathbf{r}_1, \mathbf{r}_2) \psi_k(\mathbf{r}_2) \psi_l(\mathbf{r}_2) d\tau_1 d\tau_2$$

This sum appears at first sight to contain 6^4 terms, in our very simple example, and the problem of calculating two-electron integrals over LCAO orbitals in terms of those involving the basis functions is again one of the major problems in this kind of work. The beauty of the ZDO model is that most of the basis function terms in the sum vanish, and the calculation is pretty simple.

The BASIC program given as an Appendix does all this, and solves the resultant 9×9 CI matrix to give a better description of the excited states.

The CI description of the first excited state is $\Psi(\text{first}) = -0.8288\Psi(3 \rightarrow 4) + 0.5570\Psi(2 \rightarrow 5) + 0.0203\Psi(3 \rightarrow 6) - 0.0489\Psi(1 \rightarrow 5)$ and the excitation energy is 4.984 eV , which has to be compared to the experimental value of 4.90 eV. The first strong singlet $\pi \rightarrow \pi^*$ transition is at 6.94 eV, which is to be compared to my prediction of 6.897 eV. Not bad, but remember that in the visible region the human eye can distinguish colours with a 3 nm separation. So if you are predicting the colour of a dyestuff in the visible region, look very carefully at your prediction.

Notice that I didn't make any appeal to molecular symmetry in this calculation. You will probably know why $\Psi(3 \rightarrow 5)$ didn't appear in my description of the first excited state; I didn't exclude it because it had a small coefficient; I excluded it because it makes no contribution **by symmetry**. I could, of course, have saved myself some trouble by excluding this excited state from the CI calculation and by doing the group theory in order to calculate which excited state could, in principle, contribute to the excited states of each symmetry. More sophisticated calculations have to take this into account.

7.2 BUT WHICH χ_i ARE THEY?

Looking back, I seem to have made two contradictory statements about the atomic orbital basis functions χ_i used in this model. I have appealed to your chemical intuition and prior knowledge by suggesting that the χ_i should be regarded as Slater type atomic orbitals. Then again, I have told you that the atomic orbitals used in semi-empirical theories are orthogonal, $\int \chi_i(\underline{r})\chi_j(\underline{r})d\tau = 0$ if $i \neq j$. How do we reconcile these two views?

Think of ethene, where we used the simple Hückel π-electron model to describe the conjugated system. We took $2p_\pi$ atomic orbitals, one on either carbon centre. Call these atomic orbitals χ_A and χ_B, and write them as a column vector $(\chi_A, \chi_B)^T$. The matrix of overlap integrals is

$$\underline{S} = \begin{pmatrix} 1 & p \\ p & 1 \end{pmatrix}$$

where the overlap integral between the values of χ is p. The eigenvalues of this matrix are $1 \pm p$, and the normalized eigenvectors of $\underline{S}$ are given by the column vectors $\underline{v}_1 = \sqrt{(\frac{1}{2})}(1, 1)^T$ and $\underline{v}_2 = \sqrt{(\frac{1}{2})}(1, -1)^T$. A little matrix algebra will show that

$$\underline{S} = (1 + p)\underline{v}_1\underline{v}_1^T + (1 - p)\underline{v}_1\underline{v}_2^T$$

and I will tell you that powers of certain matrices can be defined in this way. The inverse of $\underline{S}$ can be defined as

$$\underline{S}^{-1} = (1 + p)^{-1}\underline{v}_1\underline{v}_1^T + (1 - p)^{-1}\underline{v}_1\underline{v}_2^T$$

(provided that $p > 1$) and the negative square root

$$\underline{S}^{-\frac{1}{2}} = (1+p)^{-\frac{1}{2}} \underline{v}_1 \underline{v}_1^T + (1-p)^{-\frac{1}{2}} \underline{v}_1 \underline{v}_2^T$$

To cut a long story short, we regard π-electron calculations as being made formally with an orthogonal atomic orbital basis set $\chi_1^{orth}, , \ldots \chi_n^{orth}$ related to the 'ordinary' Slater type orbitals $\chi_1, \chi_2, \ldots, \chi_n$ by the matrix transformation

$$(\chi_1^{orth}, \chi_2^{orth}, \ldots \chi_n^{orth}) = (\chi_1, \chi_2, \ldots \chi_n) \underline{S}^{-\frac{1}{2}}$$

(I will leave you to prove that these new atomic orbitals are orthogonal.)

The evidence for my original assertion is that, if we make a calculation for the two-electron integrals over Slater atomic orbitals for a simple conjugated molecules, and then transform the integrals to the orthogonalized set, two-electron integrals calculated over the new, orthogonalized atomic orbitals do indeed almost satisfy the ZDO approximation. The **numerical values** resulting from this transformation are not the same as the numerical values adopted in the semi-empirical theories, but that is another story.

7.3 THE 'ALL VALENCE ELECTRON' DIFFERENTIAL OVERLAP MODELS

Prior to 1965, all we had in our armoury were the Hückel theories and a very small number of rigorous calculations designated *ab initio*', to be discussed later. The aims of theoretical chemistry in those days were to give total energies and charge distributions, whilst 'practical' chemists were more concerned with enthalpy changes, reaction paths and so on. It should come as no surprise to learn that the 'real' chemists, therefore, viewed theoretical chemists with a certain scepticism. Theoreticians did little to help their case by proposing yet more complicated and obviously incorrect parameter schemes; for example, back to ethene. We normally choose to call the axis perpendicular to the molecular plane the 'z' axis. The molecule doesn't care, it has the same total energy, electric dipole moment and so on irrespective of how we make the choice of axis system.

I have to tell you that some of the more esoteric versions of extended Hückel theory did not satisfy this simple criterion! It proved possible to calculate different total energies, and even atomic populations, depending on the axis system chosen (arbitrarily) in order to discuss the molecular wavefunction.

Pople *et al* seem to have been the first authors to give a systematic treatment of this problem. They identified two important types of transformations amongst the atomic orbital basis set;

(i) Transformations which mix those orbitals on the same atom that have the same principal and azimuthal quantum numbers. For example, a rotation of axes would mix together the three 2p orbitals on a carbon atom.

(ii) Transformations which can mix any of the atomic orbitals on the same atom. For example, descriptive organic chemistry emphasizes the carbon sp^3 hybrid atomic orbitals rather than the usual 2s, 2p$_x$, 2p$_y$ and 2p$_z$ set.

In either case, the physical properties predicted by theory have to be invariant.

Pople *et al* proceeded to examine the consequences of this conclusion in the context of the ZDO approximations to all valence electron theories.

7.3.1 CNDO/1 and CNDO/2

The most elementary theory retaining the main features of electron repulsion is the complete neglect of differential overlap (CNDO) model introduced by Pople, Santry and Segal. Only valence electrons were treated explicitly, the inner shells being taken to be a part of the 'core' that modifies the nuclear potential. The zero differential overlap approximation is applied to two-electron integrals of the form

$$\int\int \chi_i(\mathbf{r}_1)\chi_j(\mathbf{r}_1)g(\mathbf{r}_1,\mathbf{r}_2)\chi_k(\mathbf{r}_2)\chi_l(\mathbf{r}_2)\mathrm{d}\tau_1\mathrm{d}\tau_2$$

which is zero unless $i = j$ and $k = l$. This leaves χ_i (centred on atom A) and $\dot{\chi}_k$ (centred on atom B) contributing. But now some of the atomic orbitals could be on the same centre. For example, we would use the 2s and all 2p orbitals of a carbon atom. In order to satisfy the rotational invariance, such an integral

$$\gamma_{AB} = \int\int \chi_i^2(\mathbf{r}_1)g(\mathbf{r}_1,\mathbf{r}_2)\chi_k^2(\mathbf{r}_2)\mathrm{d}\tau_1\mathrm{d}\tau_2$$

is taken to depend only on the nature of atoms A and B, and this is the essence of the complete neglect of differential overlap.

Once again, the emphasis was on HF calculations, so the next step in the model is to investigate the one-electron terms which are of the type

$$\int \chi_i(\mathbf{r}_1)h(\mathbf{r}_1)\chi_j(\mathbf{r}_1)\mathrm{d}\tau_1 = -\frac{h^2}{8\pi^2 m}\int \chi_i(\mathbf{r}_1)\nabla_1^2\chi_j(\mathbf{r}_1)\mathrm{d}\tau_1$$
$$-\frac{e^2}{4\pi\epsilon_0}\int \chi_i(\mathbf{r}_1)\sum_{\alpha=1}^{NUC}\frac{Z_\alpha}{R_{\alpha,1}}\chi_j(\mathbf{r}_1)\mathrm{d}\tau_1$$

The difference from π-electron theories is that several of the χ values could be centred on the same atom.

The diagonal terms are conveniently divided into those contributions which refer to one centre (nucleus A), and those which refer to the remaining centres. Atomic orbital χ_i is centred on nucleus A.

$$\int \chi_i(\mathbf{r}_1)h(\mathbf{r}_1)\chi_i(\mathbf{r}_1)\mathrm{d}\tau_1 = -\frac{h^2}{8\pi^2 m}\int \chi_i(\mathbf{r}_1)\nabla_1^2\chi_i(\mathbf{r}_1)\mathrm{d}\tau_1$$
$$-\frac{e^2}{4\pi\epsilon_0}\int \chi_i(\mathbf{r}_1)\sum_{\alpha=1}^{NUC}\frac{Z_\alpha}{R_{\alpha,1}}\chi_i(\mathbf{r}_1)\mathrm{d}\tau_1$$

For the one-centre contribution, we note that

$$\int \chi_i(\mathbf{r}_1) h(\mathbf{r}_1) \chi_i(\mathbf{r}_1) d\tau_1 = -\frac{h^2}{8\pi^2 m} \int \chi_i(\mathbf{r}_1) \nabla_1^2 \chi_i(\mathbf{r}_1) d\tau_1$$

$$-\frac{e^2}{4\pi\epsilon_0} \int \chi_i(\mathbf{r}_1) \frac{Z_A}{R_{A,1}} \chi_i(\mathbf{r}_1) d\tau_1$$

is obviously an atomic quantity; it is written U_{ii} and obtained semi-empirically from atomic spectroscopic data.

The remaining terms

$$-\frac{e^2}{4\pi\epsilon_0} \int \chi_i(\mathbf{r}_1) \sum_{\beta \neq A} \frac{Z_\beta}{R_{\beta,1}} \chi_i(\mathbf{r}_1) d\tau_1$$

represent physically the interaction of the electron in atomic orbital χ_i with the remaining nuclear cores. We write each of these terms as $-V_{AB}$, and refer to them as penetration integrals.

The off-diagonal terms are treated to a similar analysis. First,

$$-\frac{e^2}{4\pi\epsilon_0} \int \chi_i(\mathbf{r}_1) \sum_{\beta \neq A} \frac{Z_\beta}{R_{\beta,1}} \chi_j(\mathbf{r}_1) d\tau_1$$

Each penetration term involving two orbitals on the same nuclear centre is also given the value $-V_{AB}$, in order to preserve the spirit of the ZDO approximations. Suppose now that χ_i is centred on nucleus A and χ_j on nucleus B. That leaves terms like these.

$$\int \chi_i(\mathbf{r}_1) h(\mathbf{r}_1) \chi_j(\mathbf{r}_1) d\tau_1 = -\frac{h^2}{8\pi^2 m} \int \chi_i(\mathbf{r}_1) \nabla_1^2 \chi_j(\mathbf{r}_1) d\tau_1$$

$$-\frac{e^2}{4\pi\epsilon_0} \int \chi_i(\mathbf{r}_1) \frac{Z_A}{R_{A,1}} \chi_j(\mathbf{r}_1) d\tau_1 - \frac{e^2}{4\pi\epsilon_0} \int \chi_i(\mathbf{r}_1) \frac{Z_B}{R_{B,1}} \chi_j d\tau_1$$

$$-\frac{e^2}{4\pi\epsilon_0} \int \chi_i(\mathbf{r}_1) \sum_{\alpha \neq A,B}^{NUC} \frac{Z_\alpha}{R_{\alpha,1}} \chi_j(\mathbf{r}_1) d\tau_1$$

We neglect the three-centre nuclear attraction terms but retain the two centre ones, because these are the integrals that take account of the basic bonding capacity.

To summarize then, a CNDO all-valence electron SCF calculation on a molecule such as water would, therefore, involve calculating the Hartree Fock matrix with elements

$$h_{ii}^{\mathrm{F}} = U_{ii} + \left(P_{1,AA} - \frac{1}{2} P_{1,ii}\right)\gamma_{AA} + \sum_{B \neq A}(P_{1,BB}\gamma_{AB} - V_{AB})$$

$$h_{ij}^{\mathrm{F}} = \beta_{AB}^0 S_{ij} - \frac{1}{2} P_{1,ij}\gamma_{AB}$$

(where I have collected some terms together for the sake of neatness) and repeating the iterations until the charge density matrix was self consistent. Orbital χ_i is taken to be centred on nucleus A, and χ_j on nucleus B. The expressions are correct if $A = B$, and the $P_{1,AA}$ term is found by adding together all the P_1 values for every orbital on the given atomic centre A.

The original parameter scheme was called CNDO/1 (Pople and Segal 1965), and allowed treatments of hydrogen and all first row atoms except helium. The atomic orbital basis set was taken as Slater type orbitals (STO), and an exponent of 1.2 was generally used for the hydrogen atoms.

Overlap integrals are calculated explicitly, but the atomic orbital basis functions are assumed to be orthogonal for the purposes of the HF calculation.

Electron repulsion integrals γ_{AB} are calculated explicitly using the appropriate formulae for atomic s orbitals, and the V_{AB} penetration integral was calculated explicitly, again assuming s orbitals on each centre. The U values were found from observed atomic orbital energy levels and from the γ_{AA} already discussed. Finally, that leaves the 'bonding' parameters β^0_{AB}. In order to keep the number of parameters manageable, these were taken to depend only on the contributing atoms, and a very simple additive scheme was adopted

$$\beta^0_{AB} = \beta^0_A + \beta^0_B$$

In an interesting variation on the usual theme, the β values were actually chosen by comparing the CNDO HF matrix elements with corresponding values from non-empirical calculations. This idea was novel.

It turned out that CNDO/1 calculations gave poor predictions for molecular equilibrium geometries, and this failing was analysed as being due to the treatment of the penetration terms and the U values.

The CNDO/2 parameterization came next (Pople and Segal 1966); it differs from the CNDO/1 model in the way that it handles the values of V_{AB} and the U_i; the V_{AB} is no longer calculated analytically, but just taken as $Z_B\gamma_{AB}$, where Z_B is the effective nuclear charge (in atomic units) of nucleus B. The U_i terms become

$$U_i = -\frac{1}{2}(I_i + A_i) - (Z_A - \frac{1}{2})\gamma_{AA}$$

where I_i is the ionization energy of an appropriate valence state atomic orbital and A_i the corresponding electron affinity. The basic SCF LCAO equations for CNDO/2 are

$$h^F_{ij} = -\frac{1}{2}(I_i + A_i) + ((P_{1,AA} - Z_A) - \frac{1}{2}(P_{1,ii} - 1))\gamma_{AA} + \sum_{B \neq A}(P_{1,BB} - Z_B)\gamma_{AB}$$

$$h^F_{ij} = \beta^0_{AB}S_{ij} - \frac{1}{2}P_{1,ij}\gamma_{AB}$$

At this point it is wise to study an example, and I will take H_2O with a bond length of 0.956 Å and a bond angle of 104.5. If we take the molecule to lie in the yz plane, and the centre of coordinates to be the centre of mass of the molecule, then the cartesian coordinates are given in Table 7.3.

Table 7.3 H_2O CNDO/2 calculation

Atom	x (Å)	y (Å)	z (Å)
O	0	0	0.1171
H1	0	0.7559	−0.4682
H2	0	−0.7599	−0.4682

The repulsion integrals γ_{AB} are easily calculated, assuming STOs on each centre (with a hydrogen exponent of 1.2) and are given in Table 7.4.

Table 7.4   Repulsion integrals γ_{AB} (E_h)

	O	H1	H2
O	0.8265	0.5013	0.5013
H1		0.7500	0.3422
H2			0.7500

The overlap integrals (Table 7.5) can also be calculated quite simply from the master formulae of Mulliken et al.

Table 7.5   Overlap integrals for H_2O

	O 2s	O $2p_x$	O $2p_y$	O $2p_z$	H1 1s	H2 1s
O 2s	1	0	0	0	0.4803	0.4803
O $2p_x$		1	0	0	0	0
O $2p_y$			1	0	0.3031	−0.3031
O $2p_z$				1	−0.2347	−0.2347
H1 1s					1	0.2708
H2 1s						1

You probably have access to a package that will do CNDO/2 calculations for such a straightforward molecule, and you will find the LCAO-MO coefficients and orbital energies given in Table 7.6. In any case, I have included one on your disk.

Table 7.6   Orbital energies ϵ and LCAO coefficients; CNDO/2 on H_2O

ϵ_i/E_h	O 2s	O $2p_x$	O $2p_y$	O $2p_z$	H1 1s	H2 1s
−1.488	−0.8617	0	0	0.0696	−0.3554	−0.3554
−0.788	0	0	0.7589	0	0.4605	−0.4605
−0.713	−0.3164	0	0	−0.8480	0.3006	0.3006
−0.655	0	1	0	0	0	0
0.336	−0.3966	0	0	0.5254	0.5323	0.5323
0.351	0	0	−0.6512	0	0.5366	−0.5366

7.3.2 CNDO/S

As I hinted earlier, a major objective for practitioners of ZDO π-electron theories was the treatment of molecular electronic spectra. These calculations were almost always made at the 'SCF + singly excited CI' level of theory.

Users of CNDO theories had different goals, usually the prediction of ground state electronic properties, molecular geometries and reaction mechanisms, atomization energies, etc. When CNDO theories were applied to the problem of electronic spectra, it was found that both CNDO/1 and CNDO/2 tended to vastly overestimate transition energies and they often predicted the wrong ordering for excited states. Something went wrong somewhere!

The problem of interest was the treatment of both $\pi \rightarrow \pi^*$ and $n \rightarrow \pi^*$ transitions in molecules such as pyridine, where π-electron theories cannot explain the $n \rightarrow \pi^*$ transitions. Probably the most successful treatment of excited states within the CNDO approach is that of Del Bene and Jaffé, who made three modifications to the original parameterization. The first two modifications were really just minor tinkering with the integral evaluation: the key point in the theory was their treatment of the β parameters

$$\beta_{ij}^{\sigma} = \frac{1}{2}(\beta_A^0 + \beta_B^0)S_{ij}$$

$$\beta_{ij}^{\pi} = \frac{1}{2}(\beta_A^0 + \beta_B^0)S_{ij}\,\kappa$$

Use of the CNDO Method in Spectroscopy: I Benzene, Pyridine and the Diazines

Janet Del Bene and H. H. Jaffé

Journal of Chemical Physics, **48** (1968) 1807–1813

The CNDO method has been modified by substitution of semiempirical Coulomb integrals similar to those used in the Pariser–Parr–Pople method, and by the introduction of a new empirical parameter to differentiate resonance integrals between σ orbitals and π orbitals. The CNDO method with this change in parameterization is extended to the calculation of electronic spectra and applied to the isoelectronic compounds benzene, pyridine, pyridazine, pyrimidine and pyrazine. The results obtained were refined by a limited CI calculation, and compared with the best available experimental data. It was found that the agreement was quite satisfactory for both $n \rightarrow \pi$ and $\pi \rightarrow \pi^*$ singlet transitions. The relative energies of the π and the lone pair orbitals in pyridine and the diazines are compared and an explanation proposed for the observed orders. Also, the nature of the 'lone pairs' in these compounds is discussed.

---

depending whether the pair of atomic orbitals χ_i and χ_j were locally 'σ' or 'π' type in their orientation to each other. Variation of κ over a range of possible values showed that $\kappa = 0.585$ gives the best agreement with spectroscopic data for such conjugated molecules. Apart from that, the authors also adjusted the original β_A^0 values.

Whilst this produced a sound treatment of the $n \to \pi^*$ transitions, the results of the $\pi \to \pi^*$ transitions were really no better than those given in ZDO π-electron theories.

7.3.3 INDO

In schemes such as CNDO, a typical repulsion integral

$$\int\int \chi_i(\mathbf{r}_1)\chi_j(\mathbf{r}_1)g(\mathbf{r}_1,\mathbf{r}_2)\chi_k(\mathbf{r}_2)\chi_l(\mathbf{r}_2)d\tau_1 d\tau_2$$

is zero unless $i = j$ and $k = l$. One of the major limitations of such schemes is the exclusion of one-centre repulsion integrals of the type

$$\int\int \chi_i(\mathbf{r}_1)\chi_j(\mathbf{r}_1)g(\mathbf{r}_1,\mathbf{r}_2)\chi_i(\mathbf{r}_2)\chi_j(\mathbf{r}_2)d\tau_1 d\tau_2$$

where χ_i and χ_j are on the same atom. Thus for example, CNDO theory would not distinguish energetically between the ^3P, ^1D and ^1S states arising from the orbital configuration $(1s)^2(2s)^2(2p)^2$, and would also give zero σ spin density in a planar π radical such as CH_3. Pople, Beveridge and Dobosh introduced INDO in 1967. INDO is CNDO/2 with a more realistic treatment of one-centre integrals, and in the spirit of such methods the non-zero integrals were chosen by appeal to atomic experimental spectral data. Retention of these two-electron integrals also changes the values used for the U_{ii} terms discussed above. The authors concluded that, although INDO was little better than CNDO/2 for predicting geometries, it gave a better description of singlet-triplet splitting and a much better description of spin densities.

Just to finish this train of thought, the next development ought to be to retain all one centre integrals of the type

$$\int\int \chi_i(\mathbf{r}_1)\chi_j(\mathbf{r}_1)g(\mathbf{r}_1,\mathbf{r}_2)\chi_k(\mathbf{r}_2)\chi_l(\mathbf{r}_2)d\tau_1 d\tau_2$$

where all the orbitals refer to the same atomic centre. This is the essence of NDDO, the neglect of diatomic differential overlap. The NDDO method had the advantage that it treated lone pair effects, but it never really caught on.

7.3.4 MINDO

At this point, enter stage left Dewar and the modified intermediate neglect of differential overlap (MINDO). Dewar's objective was to develop a parameterization scheme that would be a genuine working model for experimentalists. This is a very worthwhile goal, but with certain implications. The model has to be capable of giving results within chemical accuracy for a range of molecular properties, interactions and reactions. Obviously, in the context of a semi-empirical model, it has to be parameterized for a chemically useful range of atoms and not just C/H/O. It has to be implementable on a wide range of computer platforms, so there is very little point in writing the code in an obscure or non-standard language. Finally, it has to deliver the goods at reasonable cost.

I am conscious of having missed many sets of acronyms from my guided tour of differential overlap models, and I will just tell you that MINDO, MINDO/1 and MINDO/2 have disappeared into oblivion. With MINDO/3, Dewar thought that at last, he had developed a reliable model. The abstract to the landmark paper is terse;

Ground States of Molecules. XXV. MINDO/3. An Improved Version of the MINDO Semiempirical SCF-MO Method
Richard C. Bingham, Michael J. S. Dewar and Donald H. Lo
Journal of the American Chemical Society, **97** (1975) 1285–1293

The problems involved in attempts to develop quantitative treatments of organic chemistry are discussed. An improved version (MINDO/3) of the MINDO semiempirical SCF-MO treatment is described. Results obtained for a large number of molecules are summarized.

---

but the 'Summary and Conclusions' section is more revealing.

Summary and Conclusions

MINDO/3 has thus proved to be an extraordinarily versatile procedure, giving good results for every ground state property so far studied and apparently offering hope of equal extension to excited states and photochemistry. Whilst it has not yet achieved 'chemical' accuracy, the average error in the calculated heats of atomization being *ca* $6 \, \text{kCal mol}^{-1}$ instead of $1 \, \text{kCal mol}^{-1}$, it has given no unreasonable results except in one area where the MINDO approximation would be expected to fail. The errors in the heats of atomization are in any case less by orders of magnitude than those (given by non-empirical) calculations or other semiempirical MO procedures. Moreover the results for activation energies of reactions and for heats of formation of 'non-classical' ions seem definitely superior to those from (non-empirical) calculations, although admittedly there are few examples for comparison since few meaningful (non-empirical) calculations have been reported.

The basic HF equations for MINDO/3 are as follows

$$h_{ii}^{\text{F}} = U_{ii} - \sum_{B \neq A} Z_B \gamma_{AB} + \sum_{j(\text{on}A)} \left(P_{1,jj} G_{ij} - \frac{1}{2} P_{1,ij} H_{ij} \right) + \sum_{B \neq A} P_{1,BB} \gamma_{AB}$$

$$h_{ij}^{\text{F}} = \beta_{AB} S_{ij} - \frac{1}{2} P_{1,ij} \gamma_{AB}$$

where G_{ij} is a one-centre two-electron coulomb integral involving orbitals on the same atomic centre

$$G_{ij} = \int\int \chi_i(\mathbf{r}_1)\chi_i(\mathbf{r}_1)g(\mathbf{r}_1,\mathbf{r}_2)\chi_j(\mathbf{r}_2)\chi_j(\mathbf{r}_2)d\tau_1 d\tau_2$$

and H_{ij} is the corresponding exchange integral, again involving orbitals on the same atomic centre

$$H_{ij} = \int\int \chi_i(\mathbf{r}_1)\chi_j(\mathbf{r}_1)g(\mathbf{r}_1,\mathbf{r}_2)\chi_i(\mathbf{r}_2)\chi_j(\mathbf{r}_2)d\tau_1 d\tau_2$$

The core–core interactions were also changed, and in MINDO/3 these terms were made a function of the two-electron repulsion integrals.

So, instead of taking the repulsion between the two nuclei A and B with effective nuclear charges Z_A and Z_B to be

$$\epsilon_{core,AB} = \frac{e^2}{4\pi\epsilon_0}\frac{Z_A Z_B}{R_{AB}}$$

we take

$$\epsilon_{core,AB}/\frac{e^2}{4\pi\epsilon_0} = (1-a)Z_A Z_B \gamma_{AB} + aZ_A Z_B/R_{AB}$$

where a is an exponential function of the distance.

7.3.5 MNDO

The modified neglect of a differential overlap (MNDO) model should probably come next; MNDO is like INDO except that MNDO treats the diatomic repulsion integrals more accurately. It retains all two-centre terms involving 'monatomic' differential overlap, and the paper to read is probably

Ground States of Molecules. 39. MNDO Results for Molecules Containing Hydrogen, Carbon, Nitrogen and Oxygen
Michael J. S. Dewar and Walter Thiel
Journal of the American Chemical Society, **99** (1977) 4907–4917

Heats of formation, molecular geometries, ionization potentials, and dipole moments are calculated by the MNDO method for a large number of molecules. The MNDO results are compared with the corresponding MINDO/3 results on a statistical basis. For the properties investigated, the mean absolute errors in MNDO are uniformly smaller than those in MINDO/3 by a factor of about 2. Major improvements of MNDO over MINDO/3 are found for heats of formation of unsaturated systems and molecules with NN bonds, for bond angles, for higher ionization potentials, and for dipole moments of compounds with heteroatoms.

So finally, what is state-of-the-art, and do semi-empirical models have a future? The former question first, which brings me to the current scenario (1994).

7.3.6 AM1

Again, let Dewar give the summary of the Austin Model 1 (AM1).

AM1: A New General Purpose Quantum Mechanical Molecular Model
Michael J. S. Dewar, Eve G. Zoebisch, Eamonn F. Healey and James J. P.
Stewart
Journal of the American Chemical Society, **107** (1985) 3902–3909

A new parametric quantum mechanical model, AM1 (Austin Model 1), based on the NDDO approximation, is described. In it the major weakness of MNDO, in particular failure to reproduce hydrogen bonds, have been overcome without any increase in computer time. Results for 167 molecules are reported. Parameters are currently available for C, H, O and N.

According to Dewar, the errors in MNDO had a common tendency to overestimate repulsion between atoms at their van der Waals distances, and attempts to rectify this bad behaviour usually involved changes to the core repulsion integral. In AM1 the core repulsion integrals are modified by multiplying them by a sum of Gaussian functions, whose expansion coefficients and exponents were all optimized. Dewar concludes 'The main gains are the ability of AM1 to reproduce hydrogen bonds and the promise of better estimates of activation energies for reactions.' At the time of writing (September 1994), AM1 had been parameterized for H, B, C, Si, N, O, S, F, Cl, Br, I, Hg and Zn. Several molecular modelling packages such as MOBY 1.5 include AM1 and MNDO as simple semi-empirical quantum models.

7.3.7 PM3

The final word (1995) seems to be PM3, the parametric model number 3. This is the third parameterization of MNDO, with AM1 being the second. The driving force is Stewart (1989).

7.4   IS THERE A FUTURE FOR SEMI-EMPIRICAL CALCULATIONS?

The answer must, obviously, be a resounding 'yes'. There are two very divergent views on the *raison d'etre* for models in chemistry. The first view is that modelling seeks absolute truth and, therefore, all calculations have to be made as accurately as possible, to the last possible significant figure. This limits the scope of investigations to LiH.

The counter view is that modelling is a useful tool in the armoury of a practical chemist. The idea of semi-empirical models is that they should be capable of prediction, which they obviously are. Semi-empirical models will continue to play a major role in our understanding of chemistry. Whether or not you get your papers published is another matter!

Semi-empirical models come into their own when you are trying to

- correlate the properties of a number of related compounds
- cautiously extrapolate molecular properties from existing knowledge
- tackle problems that are interesting but simply outside the reach of more rigorous models.

8 Atomic Orbital Choice

But which atomic orbitals should we use in a molecular calculation?

8.1 HYDROGENIC ORBITALS

This question is as old as quantum chemistry, so let's go back to the beginning. The electronic Schrödinger equation for a one-electron atom such as hydrogen or He^+ is

$$-\frac{h^2}{8\pi^2\mu}\nabla^2\psi_{el} + (\epsilon_{el} - U)\psi_{el} = 0$$

where μ is the reduced mass of an electron and the nucleus, the nuclear charge is Ze and U is the electrostatic mutual potential energy of the electron and the nucleus, $-Ze^2/(4\pi\epsilon_0 r)$.

The electronic Schrödinger problem is identical in principle with that of the hydrogen atom, and the solutions are well known to be

$$\psi_{el} = R_{nl}(r)\, Y_{l,\pm m}(\theta, \phi)$$

where the Y's are angular momentum eigenfunctions. The energies turn out to be

$$\epsilon_n = -\frac{\mu e^4 Z^2}{8\pi^2\epsilon_0^2 n^2}$$

and these energies are essentially in perfect agreement with experiment. The energy only depends on n, the principal quantum number, and it is usual to work with the so-called **real equivalent** orbitals formed as linear combinations of the degenerate pairs having quantum numbers $+m$ and $-m$.

The first few unnormalized solutions are shown in Table 8.1, and I have used the symbol $\rho = Zr/a_0$.

Table 8.1

n	l	m	ψ_{el}
1	0	0	$\exp(-\rho)$
2	0	0	$(2 - \rho)\exp(-\rho/2)$
2	1	0	$\rho\exp(-\rho/2)\cos\theta$
2	1	± 1	$\exp(-\rho/2)\sin\cos\phi$ or $\sin\phi$

If we write the exponential as $\exp(-(Z/n)r/a_0)$, we see that it is of the form $\exp(-\zeta r/a_0)$. We refer to ζ as the **orbital exponent**. Likewise, note that $z = r\cos\theta$, and the solutions are often referred to as $p_z, \ldots$ atomic orbitals.

8.2 SLATER'S RULES

Exact solutions to the Schrödinger equation are not possible for many-electron atoms, but atomic SCF calculations have been made using such a set of atomic orbitals and treating the orbital exponents as variational parameters; what happens is that the atomic SCF calculation is done again and again, for different values of the orbital exponents, and then that set of exponents which gives the lowest energy is chosen. What might we expect?

A polyelectron atom is different from a one-electron atom because of electron shielding. The outer electron in Li will 'see' an effective nucleus with a charge rather less than $3e$ (but greater than $1e$), because of the shielding of the outer electron by the $1s^2$ core. Slater seems to have been the first person to give a set of rules for finding the orbital exponents (**without** the benefit of atomic SCF calculations). His rules are still widely quoted (Slater 1930) and are listed as follows.

Slater's Rules (1930)

1. $\zeta = (Z - S)/n$ where Ze is the nuclear charge and S (the screening constant) represents the screening of this electron by the other electrons in the atom.
2. The orbitals are divided into the following groups: 1s;2s, 2p;3s, 3p;3d; etc
3. S is calculated as a sum as follows
    (i)   Nothing from any electron in a group higher than the one considered
    (ii)   An amount 0.35 for every other electron in the group (except for the 1s group, where 0.30 is taken).
    (iii)   For an sp group, an amount 0.85 from each electron in the next lower group and 1.00 for all electrons nearer the nucleus. For a d group, an amount 1.00 for every inner electron.

In his honour, we refer to such atomic orbitals as **Slater type orbitals** (or STOs).

8.3 CLEMENTI AND RAIMONDI

Clementi and Raimondi actually **did** the atomic SCF calculations, which they reported in their classic study.

Atomic Screening Constants from SCF Functions
E. Clementi and D. L. Raimondi
Journal of Chemical Physics, **38** (1963) 2686–2689

The self consistent field function for atoms with 2 to 36 electrons are computed with a minimal basis set of Slater-type orbitals. The orbital exponent of the atomic orbitals are optimized as to ensure the energy minimum. The analysis of the optimized orbital exponents allows us to obtain simple and accurate rules for the 1s, 2s, 3s, 4s, 2p, 3p, 4p and 3d electronic screening constants. These rules are compared with those proposed by Slater and reveal the need of accounting for the screening due to the outside electrons. The analysis of the screening constants (and orbital exponents) is extended to the excited states of the ground state configuration and to the positive ions.

A selection of their results for first-row atoms is given in Table 8.2

Table 8.2

Atom	Z	ζ_{1s}	ζ_{2s}	ζ_{2p}
He	2	1.6875		
Li	3	2.6906	0.6396	
Be	4	3.6848	0.9560	
B	5	4.6795	1.2881	1.2107
C	6	5.6727	1.6083	1.5679
N	7	6.6651	1.9237	1.9170
O	8	7.6579	2.2458	2.2266
F	9	8.6501	2.5638	2.5500
Ne	10	9.6421	2.8792	2.8792

and if you take the trouble to work out what would be expected on the basis of Slater's rules, you will find that Slater's rules are very accurate.

We refer to such a set of atomic orbitals as a **minimal basis set** because we have carried over the concepts from simple atomic structure theory, and used a single atomic orbital where needed.

The next step is to expand the atomic orbital basis set, and I hinted in Chapter 3 how we should go about it. To start with we **double** the number of atomic orbitals and then redo the variational calculation. This takes account of the so-called inner and outer regions; Clementi puts it quite nicely.

Simple Basis Set for Molecular Wavefunctions Containing First- and Second-Row Atoms
E. Clementi
Journal of Chemical Physics, **40** (1964) 1944–1945

The self consistent field functions for the ground state of the first- and second-row atoms (from He to Ar) are computed with a basis set in which two Slater-type orbitals (STOs) are chosen for each atomic orbital. The reported STOs have carefully optimized orbital exponents. The total energy is not far from the accurate Hartree–Fock energy given by Clementi, Roothaan and Yoshimine for the first-row atoms and unpublished data for the second-row atoms. The obtained basis sets have sufficient flexibility to be a most useful starting set for molecular computations, as noted by Richardson. With the addition of 3d and 4f functions, the reported atomic basis sets provide a molecular basis set which duplicates quantitatively most of the chemical information derivable by the more extended basis sets needed to obtain accurate Hartree–Fock molecular functions.

We refer to such a basis set as **double zeta**. Where the minimal basis set for lithium had a 1s orbital with exponent 2.6906, the double zeta basis set has two 1s orbitals with exponents 2.4331 and 4.5177 (the inner and outer orbitals). Notice Clementi's comment about 3d and 4f functions. These atomic orbitals play no part in the description of atomic electronic ground states for first- and second-row atoms, but on molecule formation the atomic electron density distorts and such **polarization functions** are needed in order to describe the distortion. A selection from Clementi's double zeta basis set is given in Table 8.3.

So, that accounts for atoms. The general philosophy is that Slater orbitals (those which have a factor $\exp(-\zeta r/a_0)$) are the correct ones to use for accurate descriptions of atomic charge densities.

Table 8.3   A selection from Clementi's double zeta basis set

Atom	Z	1s exponents	2s exponents	2p exponents
He	2	1.4461		
		2.8622		
Li	3	2.4331	0.6714	
		4.5177	1.9781	
Be	4	3.3370	0.6040	
		5.5063	1.0118	
B	5	4.3048	0.8814	1.0037
		6.8469	1.4070	2.2086
C	6	5.2309	1.1678	1.2557
		7.9690	1.8203	2.7263
N	7	6.1186	1.3933	1.5059
		8.9384	2.2216	3.2674
O	8	7.0623	1.6271	1.6537
		10.1085	2.6216	3.6813
F	9	7.9179	1.9467	1.8454
		11.0110	3.0960	4.1710
Ne	10	8.9141	2.1839	2.0514
		12.3454	3.4921	4.6748

8.4   GAUSSIAN ORBITALS

But what about molecules?

The orbital model dominates chemistry, and at the heart of the orbital model is the LCAO self consistent field method. The main problem in these molecular structure calculations is the evaluation of integrals. Even in the LCAO SCF method, we have to evaluate a large number of integrals in order to construct the SCF Hamiltonian matrix, particularly the notorious electron repulsion integrals

$$\int\int \chi_i(\mathbf{r}_1)\chi_j(\mathbf{r}_1)g(\mathbf{r}_1,\mathbf{r}_2)\chi_k(\mathbf{r}_2)\chi_l(\mathbf{r}_2)d\tau_1 d\tau_2$$

When $i = j$ and $k = l$, this integral gives the mutual electrostatic potential energy of a pair of charges which are distributed through space, one with density $-e\chi_i^2(\mathbf{r}_1)$ the other with density $-e\chi_k^2(\mathbf{r}_2)$

They are **nasty** integrals for several reasons:

- each atomic orbital could be a Slater orbital with a factor $\exp(-\zeta r/a_0)$. The atomic orbitals will generally all be centred on different atoms, and there is no obvious choice of coordinate origin for the calculation of the integral;
- the integrals are six-dimensional;
- the integrals have a singularity; they become infinite when the electrons get close together;

- there are an awful lot of them (roughly, $n^4/8$ where n is the number of atomic orbitals).

The breakthrough came with Boys' classic paper on the use of Gaussian type orbitals (GTOs). These atomic orbitals have an exponential form of $\exp(-\alpha r^2/a_0^2)$ rather than $\exp(-\zeta r/a_0)$, but it turns out that the 'nasty' integrals are now straightforward. The reason is as follows. Consider a pair of 1s Gaussian orbitals as shown in Figure 8.1; I will call them $G_A(\mathbf{r}_A, \alpha_A)$ and $G_B(\mathbf{r}_B, \alpha_B)$. The centre of Gaussian A is at position vector $\mathbf{r}_A$, and the Gaussian exponent is α_A. An electron is situated at point $\mathbf{r}$, and so we would find a term

$$G_A(\mathbf{r} - \mathbf{r}_A, \alpha_A) = \left(\frac{2\alpha_A}{\pi}\right)^{\frac{3}{4}} \exp\left[-\alpha_A(\mathbf{r} - \mathbf{r}_A)^2\right]$$

to describe the electron 'in' G_A, and a corresponding term

$$G_A(\mathbf{r} - \mathbf{r}_B, \alpha_B) = \left(\frac{2\alpha_B}{\pi}\right)^{\frac{3}{4}} \exp\left[-\alpha_B(\mathbf{r} - \mathbf{r}_B)^2\right]$$

to describe the electron 'in' G_B.

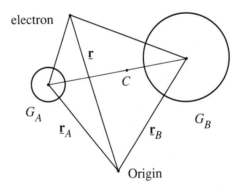

Figure 8.1   Two Gaussian orbitals G_A and G_B. C lies along the line of centres

In a variational calculation, we would have to evaluate typically one-electron integrals and two-electron integrals, which contain terms like

$$\left(\frac{2\alpha_A}{\pi}\right)^{\frac{3}{4}}\left(\frac{2\alpha_B}{\pi}\right)^{\frac{3}{4}}\int \exp\left[-\alpha_A(\mathbf{r} - \mathbf{r}_A)^2\right]\exp\left[-\alpha_B(\mathbf{r} - \mathbf{r}_B)^2\right]\ldots d\tau$$

The product of the two exponentials is equal to a third Gaussian situated at point C along the line AB, as shown in Figure 8.1. Apart from the normalizing factor, the product $G_C(\mathbf{r}, \alpha_C)$ is given by

$$G_C(\mathbf{r}_C, \alpha_C) = \exp\left(-\frac{\alpha_A \alpha_B}{\alpha_A + \alpha_B} \frac{r_{AB}^2}{a_0^2}\right) \exp\left(-\alpha_C \frac{r_C^2}{a_0^2}\right)$$

where

$$\underline{\mathbf{r}}_C = \frac{1}{\alpha_A + \alpha_B}(\alpha_A \underline{\mathbf{r}}_A + \alpha_B \underline{\mathbf{r}}_B)$$

is the centre of the product Gaussian and $\alpha_C = \alpha_A + \alpha_B$.

Apart from the normalizing factor, the two-centre integral then reduces to a simple one-centre integral of the type

$$\int \exp(-\alpha_C(\mathbf{r} - \underline{\mathbf{r}}_C)^2) \ldots d\tau$$

The simplification is more dramatic for a two-electron integral, which can involve GTOs on four different centres. The integrals can be reduced to simple formulae. Formulae for integrals involving higher symmetry types can be deduced from the s-type formulae by differentiation. Here is the famous synopsis.

Electronic Wave Functions
I. A General Method of Calculation for the Stationary States of any Molecular System
S. F. Boys
Proceedings of the Royal Society of London Series A, **200** (1950) 542–554

This communication deals with the general theory of obtaining numerical electronic wavefunctions for the stationary states of atoms and molecules. It is shown that by taking Gaussian functions, and functions derived from these by differentiation with respect to the parameters, complete systems of functions can be constructed appropriate to any molecular problem, and that all the necessary integrals can be explicitly evaluated. These can be used in connection with the molecular orbital method, or localized bond method, or the general method of treating linear combinations of many Slater determinants by the variational procedure. This general method of obtaining a sequence of solutions converging to the accurate solution is examined. It is shown that the only obstacle to the evaluation of wavefunctions of any required degree of accuracy is the labour of computation. A modification of the general method applicable to atoms is discussed and considered to be extremely practicable.

These Gaussian basis functions (GTOs) have the advantage that all the integrals needed for LCAO calculations can be done using simple formulae. It is normal to use only 1s, 2p, 3d Gaussian orbitals in molecular calculations rather than 1s, 2s, 3s, 2p, 3p... One problem is that the GTOs do not give terribly good energies. If we use a single GTO orbital for a H atom

$$G(\alpha) = \left(\frac{2\alpha}{\pi}\right)^{\frac{3}{4}} \exp\left(-\alpha\frac{r^2}{a_0^2}\right)$$

and calculate the best orbital exponent α, we find $\alpha_{opt} = 0.283$ and the variational energy is $-0.42441E_h$ (to be compared with the experimental value of $-\frac{1}{2}E_h$, an error of some 15%. A quick glance at Figure 8.2 of this GTO and the correct STO shows that GTOs also have an incorrect behaviour at the nucleus and that they also fall off too quickly as the electron gets further from the nucleus, although that is hard to spot from the graph.

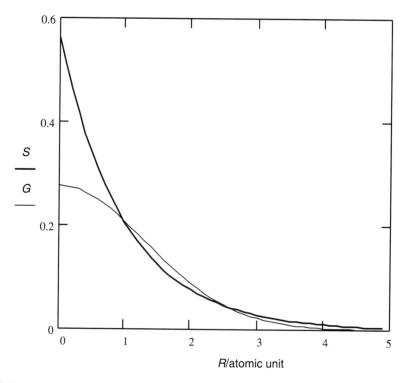

Figure 8.2 Behaviour of Slater orbital (S) and Gaussian orbital (G) with distance from the nucleus in a H atom

8.4.1 The STO/nG philosophy

The next step was to represent each Slater atomic orbital as a fixed linear combination of Gaussian orbitals; so a Slater type orbital with exponent ζ, $S(\zeta)$ is written, for example, as a sum of three **primitive** Gaussians

$$S(\zeta) = d_1 G_1(\alpha_1) + d_2 G_2(\alpha_2) + d_3 G_3(\alpha_2)$$

and the values of d and α are usually found from least-squares fitting. (Hehre *et al*, 1969).

Self Consistent Molecular-Orbital Methods
I. Use of Gaussian Expansions of Slater-Type Atomic Orbitals
W. J. Hehre, R. F. Stewart and J. A. Pople
Journal of Chemical Physics, **51** (1969) 2657–2665

Least Squares representations of Slater-type atomic orbitals as a sum of Gaussian-type orbitals are presented. These have the special feature that common Gaussian exponents are shared between Slater-type 2s and 2p functions. Use of these atomic orbitals in self consistent molecular-orbital calculations is shown to lead to values of atomization energies, atomic populations, and electric dipole moments which converge rapidly (with increasing size of Gaussian expansion) to the values appropriate for pure Slater-type orbitals. The ζ exponents (or scale factors) for the atomic orbitals which are optimized for a number of molecules are also shown to be nearly independent of the number of Gaussian functions. A standard set of ζ values for use in molecular calculations is suggested on the basis of this study and is shown to be adequate for the calculation of total and atomization energies, but less appropriate for studies of charge distribution.

We refer to such a basis set as an STO/nG basis set, and to take a concrete example, the fit for a H 1s orbital with STO exponent $\zeta = 1$ gives the family of Gaussian exponents α and STO expansion coefficients d shown in Table 8.4.

Table 8.4 Coefficients d and exponents α to fit a 1s STO ($\zeta = 1$)

	α	d
STO/2G	0.151623	0.678914
	0.851819	0.430129
STO/3G	0.109818	0.444635
	0.405771	5.35328
	2.22766	0.154329
STO/4G	0.0880187	0.291626
	0.265204	0.532846
	0.954620	0.260141
	5.21686	0.0567523
STO/5G	0.0744527	0.193572
	0.197572	0.482570
	0.578648	0.331816
	2.07173	0.113541
	11.3056	0.0221406
STO/6G	0.0651095	0.130334
	0.158088	0.416492
	0.407099	0.370563
	1.18506	0.168538
	4.23592	0.0493615
	23.31030	0.00916360

We regard each fixed linear combination as a single basis function, for the purposes of (e.g.) the SCF calculation, but for integral evaluation we must remember that it is really a fixed combination of primitive GTOs. We don't escape the necessity of calculating integrals over all the primitive orbitals. Thus, to give a STO/3G fit to a hydrogen 1s STO with STO exponent $\zeta = 1$, we need three primitive Gaussians with exponents $\alpha_1 = 0.109818$, $\alpha_2 = 0.405771$ and $\alpha_3 = 2.22766$ (Figure 8.3).

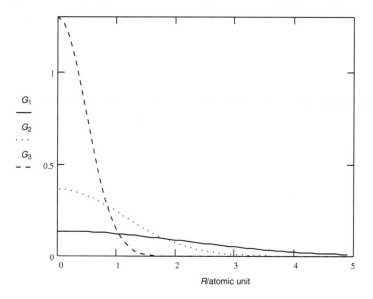

Figure 8.3   STO/3G primitives G_1, G_2 and G_3 versus distance from nucleus

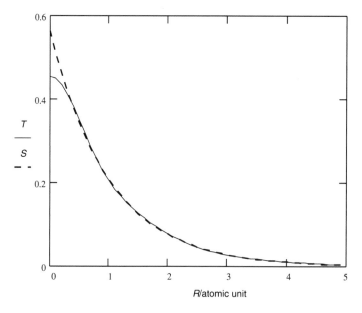

Figure 8.4   The goodness of fit between STO/3G (T) and a hydrogen 1s (S) orbital

These are then combined to give the 'contracted' basis function and, as you can see, the fit is pretty exact (apart from the bad behaviour of the Gaussian orbitals at the nucleus) (Figure 8.4).

8.4.2 The STO/4–31G story

It soon became apparent that these minimal basis sets were not particularly adequate for accurate predictions of molecular geometries because they lacked flexibility in the valence regions. The next step was to give a little more flexibility to the STO/nG basis sets by splitting the valence shell orbitals into inner and outer components, as noted by Ditchfield, Hehre and Pople.

Self-Consistent Molecular-Orbital Methods.IX An Extended Gaussian-Type Basis for Molecular-Orbital Studies of Organic Molecules
R. Ditchfield, W. J. Hehre and J. A. Pople
Journal of Chemical Physics, **54** (1971) 724–728

An extended basis set of atomic functions expresses as fixed linear combinations of Gaussian functions is presented for hydrogen and the first-row atoms carbon to fluorine. In this set, described as 4–31G, each inner shell is represented by a single basis function taken as a sum of four Gaussians and each valence orbital is split into inner and outer parts described by three and one Gaussian function, respectively. The expansion coefficients and Gaussian exponents are determined by minimizing the total calculated energy of the atomic ground state. This basis set is then used in single-determinant molecular-orbital studies of a group of small polyatomic molecules. Optimization of valence-shell scaling factors shows that considerable rescaling of atomic functions occurs in molecules, the largest effects being observed for hydrogen and carbon. However, the range of optimum scale factors for each atom is small enough to allow the selection of a standard molecular set. The use of this standard basis gives theoretical equilibrium geometries in reasonable agreement with experiment.

Thus, a hydrogen atom is represented by two basis functions, the first of which is a linear combination of three GTOs, and the other a single GTO as in Table 8.5.

Table 8.5 4–31G Hydrogen-atom basis functions

Hydrogen (^{2}s)	α	d
	13.00773	0.0334960
	1.962079	0.22472720
	0.4445290	0.8137573
	0.1219492	1

For an atomic 'STO' exponent ζ different from 1, the expansion coefficients d remain the same but the exponents have to be multiplied by $\zeta^{3/2}$.

8.4.3 Extended basis sets

A different line of attack is afforded by the work of Dunning *et al*. The idea is to start out with a reasonable set of atomic GTOs and then look for a contraction scheme that will give combinations of the primitive GTOs that are more or less constant in a molecular calculation. Once again I stress that, although we imagine a set of atomic orbitals, each of these atomic orbitals can be a combination of primitive GTOs and all the necessary integrals for a quantum mechanical calculation still have to be done over the primitives. In the case of the STO philosophy, the primitives are constrained to have the same exponents for each shell, which leads to a great saving in integral evaluation time.

Dunning concentrates on energy-optimized GTOs where each exponent has been variationally optimized in an atomic SCF calculation. The GTO basis set used is that due to Huzinaga, and consists of ten primitive GTOs and six primitive GTOs for each first row atom. For example, the oxygen atomic orbital basis set is given in Table 8.6.

Table 8.6 Dunning contraction scheme for Huzinaga's GTO basis set

18050.0	s	0.000757
2660.0		0.006066
585.7		0.032782
160.9		0.132609
51.16		0.396839
17.90		0.542572
17.90	s	0.262490
6.639		0.769828
2.077	s	1
0.7736	s	1
0.2558	s	1
49.83	p	0.016358
11.49		0.106453
3.609		0.349302
1.321		0.657183
0.4821	p	1
0.1651	p	1

The primitive GTOs with exponents 18050 through 0.2558 are 1s type, and the remainder are 2p type, as I have indicated in the table. The two most diffuse s functions (with exponents 0.7736 and 0.2558) are the principal component of the atomic 2s orbital, and they are allowed to vary freely in molecular calculations. The s primitive with exponent 2.077 turns out to make contributions to both the atomic 1s and 2s orbitals, so that is also left free. The remaining seven distinct primitive s types describe the atomic 1s orbital, and a careful examination of the ratios of their contributions to the atomic orbitals led Dunning to conclude that the primitive with exponent 17.90 could be best counted twice. Similar considerations apply to the 2p orbitals.

Gaussian Basis Functions for Use in Molecular Calculations III. Contraction of (10s6p) Atomic Basis Sets for the First-Row Atoms

T. H. Dunning, Jr

Journal of Chemical Physics, **55** (1975) 716–723

Contracted [5s3p] and [5s4p] Gaussian basis sets for the first-row atoms are derived from the (10s6p) primitive basis sets of Huzinaga. Contracted [2s] and [3s] sets for the hydrogen atom obtained from primitive sets ranging in size from (4s) to (6s) are also examined. Calculations on the water and nitrogen molecules indicate that such basis sets when augmented with suitable polarization functions should yield wavefunctions near the Hartree–Fock limit.

Take note of Dunning's notation; he writes the set of primitive GTOs (10s6p) in an obvious notation, and the contracted basis functions [5s3p].

There are dozens of GTO basis sets in the literature; the first molecular GTO calculations were carried out at the University of Leeds (UK) in the 1950s, to be followed by all the development work that went into POLYATOM. In those early days, attention was given to developing basis sets needed for calculations on organic molecules. Today, entire learned volumes have appeared detailing GTO basis sets. Some basis sets are more popular than others, and you will certainly come across the STO and the Dunning varieties.

8.5 POLARIZATION AND DIFFUSE FUNCTIONS

We are almost at the end of our story.

We would normally write the electronic configuration of a ground state carbon atom as $1s^2 2s^2 2p^2$, and a great deal of intellectual activity has gone into defining mythical 'valence states' for carbon atoms in different bonding situations. But no one would include a d orbital in the description of a ground-state carbon atom. Not only that, carbon atoms in free space have spherical symmetry.

A carbon 'atom' in a molecule is a quite different entity because its charge density distorts from spherical symmetry. To take account of the finer points of this distortion, we need very often to include d, f, . . . atomic orbitals to our basis set. Such atomic orbitals are referred to as **polarization functions** because their inclusion would allow a free atom to take account of the polarization induced by an external electric field, or by molecule formation.

The most widely used basis sets seem to be those of the STO genus, so let's see what Pople et al have to say about polarization functions.

Self-Consistent Molecular Orbital Methods. XVII. Geometries and Binding Energies of Second-Row Molecules. A Comparison of Three Basis Sets.
J. B. Collins, P. von R. Schleyer, J. S. Binkley and J. A. Pople
Journal of Chemical Physics, **64** (1976) 5142–5151

Three basis sets (minimal s-p, extended s-p and minimal s-p with d functions on the second row atoms) are used to calculate geometries and binding energies of 24 molecules containing second-row atoms. d functions are found to be essential in the description of both properties for hypervalent molecules and to be important in the calculations of two-heavy-atom bond lengths even for molecules of normal valence.

The presence of polarization functions in a heavy-atom basis set is denoted by a *, and in a hydrogen atom by a further *. Thus, a STO/4–31G** basis set means that the standard STO/4-31G basis set has been augmented with d orbitals on the heavy centres and p orbitals on the hydrogen atoms.

Sometimes it turns out that we need to include a number of polarization functions, not just one of each type, and the notation (e.g.) 4–31G (3d,2p) indicates a standard 4–31G basis set augmented with 3@d type primitives on each heavy centre (ie 1.8. d primitives per centre) and 2@p type primitives on every H atom. Not only that, but studies on anions and Rydberg states (where the excited electron(s) are far from the nuclear centres) show that extra **diffuse** valence primitives are needed in the basis set. Their presence is indicated with the symbol +. So, finally, a basis set denoted

$$6 - 31 + +G(3d, 2p)$$

means that the inner shells are represented by a fixed linear combination of six primitive Gaussians but that the valence shell is represented by an inner and outer split comprising $3 + 1$ primitive Gaussians. In addition, two sets of diffuse valence primitives have been added to the basis set, and various polarization functions on every atom. Each heavy atom has had 3@d orbitals added, and each H atom has had 2@p orbitals added.

8.6 LITERATURE SOURCES

Most *ab initio* packages have a good selection of basis sets stored internally, and these can be chosen by keywords. References to the original literature are almost always to be found in the program manuals. A comprehensive collection of basis sets is given in the book by Poirer, Kari and Csizmadia (1985).

9   An *Ab Initio* Package—GAUSSIAN92

It is traditional to divide molecular models into three broad bands depending on the degree of sophistication. There are sublevels within each band and a great deal of jargon accompanied by acronyms. Many authors speak of the 'level of theory'. The most rigorous band is the *Ab Initio* one, where we write down a Hamiltonian operator and a basis set, and then do all the calculations with full rigour as in our discussion of the H_2 molecule. We can do the calculations at SCF level, at CI level and so on; I am not implying that these calculations are 'correct' in any sense, although we would hope that they would bear some relation to reality. The phrase simply means that a (quantum mechanical) model has been stated and from that point on, everything has been done with full rigour.

Empirical models are to my mind typified by molecular mechanics and the free-electron model (which I haven't discussed). We take some simple physical model for a molecular system and generally solve the very simple equations with full rigour. In the middle are the **semi-empirical** models such as PPP theory. These models usually involve starting from *Ab Initio* and then attempting to simplify the mathematical treatment by appeal to experiment.

There are differences of opinion between authors as to which models should be classified as empirical and which are semi-empirical. I would classify Hückel π-electron theory as semi-empirical, for the following reason: Hückel theory can be shown to have a relationship to the Hartree Fock (HF) model. The diagonal elements of the HF matrix are replaced by parameters α which depend only on the nature of the contributing atom. In extended Hückel theory, the diagonal elements of the Hamiltonian matrix are replaced by simple functions of the atomic ionization energies. The division is therefore subjective and many people would classify Hückel theory as empirical. But there is a serious point to bear in mind. The greater the rigour of the calculation, the higher the computer resource needed and so the greater the cost.

Over the last 25 years, international collaboration and cooperation on a scale rarely witnessed in science has led to the development of several very sophisticated software packages for *Ab Initio* molecular electronic structure calculations. Most of them cater for users whose interests in life concern large molecules. A few are oriented to users whose intention is to obtain the most accurate wavefunction and energy for LiH. Several attempt to satisfy the needs of both groups of scientists, and I am going to refer extensively to just one such package, GAUSSIAN92.

9.1 GAUSSIAN92

GAUSSIAN92 is a logical extension of the GAUSSIAN*XY* series (where XY = year of issue) of such packages associated with the names of Pople and coworkers. The GAUSSIAN packages were originally intended for use by organic chemists, but over they years they have also proved attractive for researchers interested in small systems. The input to GAUSSIAN92 is straightforward, but I must warn you that the idea of a graphical user interface is an alien concept. Think of input on cards or a Teletype, and you have got the idea. Other packages do have excellent graphical user interfaces, and a number of so-called 'molecular modelling' packages such as MOBY 1.5 have facilities for easy transfer between MM and *Ab Initio* calculations.

At its very simplest, a package of the 1970s would consist of typically 20 000 lines of FORTRAN programs to calculate one- and two-electron integrals over Gaussian orbitals and to perform SCF calculations for closed and open shell states of molecules. In addition, programs would be available for the calculation of properties such as electric dipole moments, and also programs to aid with analysis (or 'visualization', as it is now known) of the wavefunction. We have come on apace since then, but FORTRAN is still the lingua franca and a million lines of code is not uncommon in a decent *Ab Initio* package.

More modern molecular *Ab Initio* packages contain procedures to calculate molecular geometries, force fields, molecular properties such as electric dipole polarizability and magnetizability, and have the ability to deal with calculations beyond the Hartree Fock model. They also will do empirical and semi empirical calculations, but the graphical user interface is often of the dark age. Many of you will be used to the idea of a graphical user interface such as that afforded by DTMM (Chapter 1), where you draw the molecule on screen. You are in for a shock!

Note that I omitted to say 'atoms'; the theory and calculation of atomic electronic structures is very much a specialist field, with its own software and literature. There is a deep mathematical reason for this; the Hamiltonian operator for a 'light' atom commutes with the electronic angular momenta operators and so the two have simultaneous eigenvectors. This means that the theory of atomic structure is dominated by the theory and application of angular momentum and this makes life very easy (or extremely difficult depending on your viewpoint).

9.1.1 Level of theory

The first choice in an *Ab Initio* calculation is the so-called **level of theory**. For this chapter, I will concentrate on HF calculations, and this together with my choice of basis set defines the level of theory. So, a choice of a STO/4–31G basis set together with HF means a level of theory referred to throughout the GAUSSIAN-speaking world as 'HF/4–31G'.

9.1.2 Atomic orbital basis sets

These are stored internally, or you can input your own special favourite, should you so wish. GAUSSIAN92 has the following common basis sets stored internally:

STO–3G and STO–3G*	H–Xe
STO/3–21G and STO/3–21G*	H–Cl
STO/6–21G	H–Cl
STO/4–31G	H–Ne
STO/6–31G	H–Cl
STO/6–311G	H–Ne and Na–Ar
D95V	H–Ne
D95	H–Cl
SHC	H–Cl
CEP–4G	H–Cl
CEP–31G	H–Cl
CEP–121G	H–Cl
LANL1MB	H–Bi
LANL1DZ	H–Bi

and the $+$ or $++$ diffuse functions can be added to several of the basis sets, as well as multiple polarization functions.

9.1.3 Molecular geometry

The majority of molecular structure calculations are performed within the Born-Oppenheimer approximation, which means that the nuclei are 'clamped' in position for the purpose of calculating the electronic wavefunction. This means that a molecular geometry has to be input. Many PC-based packages have a graphical user interface as I showed you in chapter 1. As an alternative, it is usually possible to work out the cartesian coordinates of each nucleus and input these directly. This might be appropriate if you wish to take the geometry from a crystallographic library such as the Brookhaven National Laboratory Protein data bank.

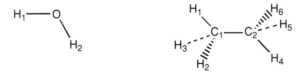

Figure 9.1 Two simple molecules for *Ab Initio* runs

At the minute, the input to many serious *Ab Initio* packages is via the so-called 'Z-matrix'. In any case, the PC-packages very often translate the screen image of a molecule into a Z-matrix, so you need to be aware of what is going on.

The idea is to define the nuclear geometry in terms of bond lengths and bond angles, etc.

```
#HF/STO-4G POP = FULL
WATER MOLECULE
 0 1
O
H1 O ROH
H2 O ROH H1 ANG

ROH  0.956
ANG  104.5
```

Let's examine a couple of simple examples of GAUSSIAN92 input, particularly the Z-matrix.

The first case is H_2O, and I have chosen a level of theory of HF/4G. The first couple of input records are therefore obvious. The next non-null record defines the charge on the molecule, and its spin multiplicity. We are dealing with an uncharged molecule (hence the zero) and an electronic state with spin multiplicity 1 (hence the 1). I find it useful at this stage to draw the molecule on a piece of paper, number the atoms and then make some educated guess as to the molecular geometry. In the case of H_2O I have taken a bond length of 95.6 pm and an bond angle of 104.5°. We then have to input the geometry which I will do using the Z-matrix technique. We start at an atom, it really doesn't matter which one, so I have chosen oxygen (labelled O which you must distinguish very carefully from the number zero 0). That is the first input record for the Z-matrix. The next Z-matrix record says that H1 is joined to O with a bond length of ROH. The third Z-matrix record says that H2 is joined to O with a bond length of ROH and that the H_1OH_2 angle is ANG. The final two records define ROH and ANG.

For a molecule having more than three atoms, we usually have to define a dihedral angle in addition. For such a molecule, then each Z-matrix record after the first three has a format

```
E N1 L N2 A N3 D
```

where, in the words of a manual 'E is the elemental symbol of the atom, N1 is a previously defined atom to which the new atom is bonded and L is the length of the bond. N2 is another atom and B the bond angle between the atom being defined (N1) and N2. N3 is another atom and D the dihedral angle between the EN1N2 and N1N2N3 planes'.

Staggered ethane (page 116) should make the process clear; I have started at C1. The next record says that C2 is joined to C1 with a bond length RCC. H1 is joined to C1 with a bond length RCH and makes an angle H1C1C2 of ANG. H2 is obviously related to H1 by a 120° rotation about the C–C bond, so the dihedral angle is +120° and H3 is related to H1 by a rotation of 120° in the opposite direction, giving a dihedral angle of −120°. To define H4, I use the fact that the dihedral angle H4C2C1H1 is 180°, and the same reasoning applies to the two remaining hydrogens. Obviously it is necessary to read the manual before starting a calculation.

There are two simple 'tricks of the trade' when writing Z-matrices. The first point concerns linear molecules. Many packages find difficulty at some stage when angles of 180° occur. These problems can be avoided by introducing dummy atoms (with symbol X in GAUSSIAN language). The example Z-matrix given here for the hydrogen-bonded complex CO . . . HF (Figure 8.2) should make the procedure clear. I have added the two dummy centres X1 and X2, with the aim of creating off-axis centres. The first dummy centre is joined to atom O, and the second dummy centre to atom H.

```
# HF/6–311G**

CO . . . HF GEOMETRY
0 1
C
O C RCO
X1 O 1. C 90.
H O ROH X1 90. C 180.
X2 H 1. O 90. X1 0.
F H RHF X2 90. O 180.

RCO 1.1306
ROH 1.9606
RHF 0.9234
```

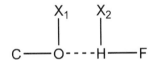

Figure 9.2   Hydrogen-bonded complex CO . . . HF

The second point concerns (particularly) cyclic structures in molecules of high symmetry. Think of furan and your first reaction would be to build up the ring starting from O, then working your way around the ring systematically until you got back to the last carbon atom joined to O. The problem is that, by the time you get to the final carbon atom, arithmetical rounding errors will have reduced the symmetry. Although the bond lengths will be more or less the same, the packages will generally recognize molecular symmetry by testing whether atoms are exactly equivalent. A difference of 10^{-5} Å will break the symmetry. You have to be careful with cyclic structures if you want to capitalize on their symmetry properties.

A solution to the problem is to define an axis of symmetry through the use of dummy atoms. For example, in furan the following Z-matrix will retain C_{2v} symmetry. What I have done is to take a dummy centre half way along the C_3–C_4, define the oxygen atom by its distance from the dummy centre and work from there. The variables in a Z-matrix do **not** have to be chemical 'bond' distances, although it has to be said that normal chemical variables are usually the best choice for geometry optimizations.

```
# HF/6-311G** FREQ

STAGGERED ETHANE GEOMETRY

0  1
C1
C2 C1 RCC
H1 C1 RCH C2 ANG
H2 C1 RCH C2 ANG H1 120.
H3 C1 RCH C2 ANG H1 -120.
H4 C2 RCH C1 ANG H1 180.
H5 C2 RCH C1 ANG H2 180.
H6 C2 RCH C1 ANG H3 180.


RCC 1.5272
RCH 1.0862
ANG 111.1884
```

```
# HF/4-31G

FURAN

0 1
X
C3 X HALF34
O X ROX C3 90.
C4 X HALF34 O 90. C3 180.
C2 C3 R23 X A23X O 0.
C5 C4 R23 X A23X O 0.
H2 C2 RCH O AA X 180.
H3 C3 RCH C2 BB H2 0.
H5 C5 RCH O AA X 180.
H4 C4 RCH C5 BB O 180.


HALF34 0.7219
ROX 2.0933
R23 1.3395
RCH 1.0632
A23X 106.676
AA 116.4592
BB 126.685
```

9.2 A GAUSSIAN92 RUN

Consider then a simple calculation at the HF/4G level of theory. Output from this particular package is just as unfriendly as the input:

```
Entering Gaussian System, Link 0 = /usr/contrib/g92/g92
Initial command:
/usr/contrib/g92/l1.exe /tmp1/tmp_g92.827/g92.int-inp -scrdir /tmp1/tmp_g92.827
Entering Link 1  =  /usr/contrib/g92/l1.exe PID = 864.
```

This software is provided under written license and may be used, copied, transmitted, or stored only in accord with that written license.

Cite this work as:
Gaussian 92, Revision E.1, M. J. Frisch, G. W. Trucks,
M. Head-Gordon, P. M. W. Gill, M. W. Wong, J. B. Foresman,
B. G. Johnson, H. B. Schlegel, M. A. Robb, E. S. Replogle,
R. Gomperts, J. L. Andres, K. Raghavachari, J. S. Binkley,
C. Gonzalez, R. L. Martin, D. J. Fox, D. J. Defrees, J. Baker,
J. J. P. Stewart, and J. A. Pople, Gaussian, Inc.,
Pittsburgh PA, 1992.

```
* * * * * * * * * * * * * * * * * * * * * * * * * * * * * * * * * * * * * * * * *
Gaussian 92: HP-PARisc-HPUX-G92RevE.1 12-Jun-1993
             12-Jul-1994
* * * * * * * * * * * * * * * * * * * * * * * * * * * * * * * * * * * * * * * * *
%save
Default route: SCF = Direct MP2 = Stingy MAXDISK = 25000000
----------------
# HF/STO-4G POP = FULL
----------------
1/29 = 10000/1;
2/12 = 2,17 = 6,18 = 5/2;
3/6 = 4,11 = 9,25 = 1,30 = 1/1,2,3;
4//1;

5/5 = 2,32 = 1,38 = 4/2;
6/7 = 3,19 = 1,28 = 1/1;
99/5 = 1,9 = 1/99;
----------------
```

Route through G92

WATER MOLECULE

Symbolic Z-matrix:
Charge = 0 Multiplicity = 1
O
H1   O   ROH
H2   O   ROH   H1   ANG
Variables:
ROH   0.956
ANG   104.5

Z-MATRIX (ANGSTROMS AND DEGREES)
CD Cent Atom N1 Length/X N2 Alpha/Y N3 Beta/Z J
----

1 1 O
2 2 H   1   .956000(1)
3 3 H   1   .956000(2)   2 104.500(3)

                Z-Matrix orientation:

Center   Atomic   Coordinates (Angstroms)
Number   Number   X   Y   Z
----

1   8   .000000   .000000   .000000
2   1   .000000   .000000   .956000
3   1   .925549   .000000   -.239363

        Distance matrix (angstroms):
            1   2   3
1 O   .000000
2 H   .956000   .000000
3 H   .956000   1.511798   .000000
        Interatomic angles:
  H2-O1-H3 = 104.5
STOICHIOMETRY   H2O
FRAMEWORK GROUP   C2V[C2(O),SGV(H2)]
DEG. OF FREEDOM   2
FULL POINT GROUP   C2V   NOP 4
LARGEST ABELIAN SUBGROUP   C2V   NOP 4
LARGEST CONCISE ABELIAN SUBGROUP   C2   NOP 2
Standard orientation:

Center   Atomic   Coordinates (Angstroms)
Number   Number   X   Y   Z

Use Molecular Symmetry
1   8   .000000   .000000   .117056
2   1   .000000   .755899   -.468224
3   1   .000000   -.755899 -.468224

Rotational constants (GHZ):   824.1771445   438.8077671   286.3496857
Isotopes: O-16,H-1,H-1
Standard basis: STO-4G (S, S=P, 5D, 7F)
There are   4 symmetry adapted basis functions of A1 symmetry.
There are   0 symmetry adapted basis functions of A2 symmetry.
There are   1 symmetry adapted basis functions of B1 symmetry.

Molecular
geometry

Use
molecular
symmetry

There are 2 symmetry adapted basis functions of B2 symmetry.
Crude estimate of integral set expansion from redundant
integrals = 1.296.
Integral buffers will be 262144 words long.
Raffenetti 1 integral format.
Two-electron integral symmetry is turned on.
    7 basis functions   28 primitive gaussians
    5 alpha electrons   5 beta electrons
     nuclear repulsion energy   9.2065546047 Hartrees.
One-electron integrals computed using PRISM.
One-electron integral symmetry used in STVInt
The smallest eigenvalue of the overlap matrix is 3.412D-01
DipDrv: MaxL = 4.
DipDrv: will hold 34 matrices at once.
PROJECTED INDO GUESS.
INITIAL GUESS ORBITAL SYMMETRIES.
   OCCUPIED (A1) (A1) (B2) (A1) (B1)
   VIRTUAL (A1) (B2)
Alpha deviation from unit magnitude is 4.44D-16 for orbital 2.
Alpha deviation from orthogonality is 1.94D-16 for orbitals 4 2.
Warning! Cutoffs for single-point calculations used.
A Direct SCF calculation will be performed.
Using DIIS extrapolation.
Closed shell SCF:
Requested convergence on RMS density matrix = 1.00D-04 within 64 cycles.
Requested convergence on MAX density matrix = 1.00D-02.
Requested convergence on   energy = 5.00D-05.
Two-electron integral symmetry used by symmetrizing Fock matrices.
Keep R1 integrals in memory in canonical form, NReq =   429938.
IEnd = 28546 IEndB =   28546 NGot =   2048000 MDV =   2027549
LenX = 2027549
MinBra = 0 MaxBra = 3 MinLOS = -1 MaxLOS = -1 MinRaf = 0 MaxRaf = 3 MinLRy = 4.
IRaf =   0 NMat =   1 IRICut =   1 DoRegI = T DoRafI = F ISym2E = 1 JSym2E = 1.
Fock matrices symmetrized in FoFDir.
Convergence on energy, delta-E = 4.12D-05
SCF DONE: E(RHF) = -75.4964937234   A.U. AFTER 4 CYCLES
     CONVG = .7012D-03   -V/T = 2.0001
     S**2 = .0000
KE = 7.549085547930D + 01 PE = -1.985352953070D + 02 EE =
3.834139149961D + 01
Copying SCF densities to generalized density rwf, ISCF = 0
IROHF = 0.

SCF
cycles

* *
    Population analysis using the SCF density.
* *

ORBITAL SYMMETRIES.
   OCCUPIED (A1) (A1) (B2) (A1) (B1)
   VIRTUAL (A1) (B2)
THE ELECTRONIC STATE IS 1-A1.
Alpha occ. eigenvalues – -20.42898 -1.27700 -.62065 -.45816 -.39622
Alpha virt. eigenvalues – .59981 .73402
   Molecular Orbital Coefficients
                 1   2   3   4   5
        (A1)–O   (A1)–O   (B2)–O   (A1)–O   (B1)–O

	EIGENVALUES —	20.42898	-1.27700	-.62065	-.45816	-.39622
1 1	O 1S	.99592	-.22414	.00000	-.09691	.00000
2	2S	.01920	.83764	.00000	.53389	.00000
3	2PX	.00000	.00000	.00000	.00000	1.00000
4	2PY	.00000	.00000	.61052	.00000	.00000
5	2PZ	-.00357	-.13022	.00000	.77784	.00000
6 2	H 1S	-.00441	.15430	.44182	-.27809	.00000
7 3	H 1S	-.00441	.15430	-.44182	-.27809	.00000

		6	7
		(A1)–V	(B2)–V
	EIGENVALUES –	.59981	.73402
1 1	O 1S	-.12337	.00000
2	2S	.88517	.00000
3	2PX	.00000	.00000
4	2PY	.00000	.98788
5	2PZ	-.74145	.00000
6 2	H 1S	-.80001	-.84070
7 3	H 1S	-.80001	.84070

DENSITY MATRIX.

		1	2	3	4	5
1 1	O 1S	2.10300				
2	2S	-.44074	1.97408			
3	2PX	.00000	.00000	2.00000		
4	2PY	.00000	.00000	.00000	.74547	
5	2PZ	-.09950	.61226	.00000	.00000	1.24400
6 2	H 1S	-.02406	-.03861	.00000	.53948	-.47277
7 3	H 1S	-.02406	-.03861	.00000	-.53948	-.47277

		6	7
6 2	H 1S	.59273	
7 3	H 1S	-.18809	.59273

Full Mulliken population analysis:

		1	2	3	4	5
1 1	O 1S	2.10300				
2	2S	-.10199	1.97408			
3	2PX	.00000	.00000	2.00000		
4	2PY	.00000	.00000	.00000	.74547	
5	2PZ	.00000	.00000	.00000	.00000	1.24400
6 2	H 1S	-.00133	-.01843	.00000	.16770	.11379
7 3	H 1S	-.00133	-.01843	.00000	.16770	.11379

			6	7
6 2	H	1S	.59273	
7 3	H	1S	-.04747	.59273

Gross orbital populations:

		1
1 1	O 1S	1.99835
2	2S	1.83523
3	2PX	2.00000
4	2PY	1.08086
5	2PZ	1.47158
6 2	H 1S	.80699
7 3	H 1S	.80699

Condensed to atoms (all electrons):

	1	2	3
1 O	7.862549	.261736	.261736
2 H	.261736	.592726	-.047473

3 H   .261736 -.047473 .592726
Total atomic charges:
                                              1
1 O   -.386022
2 H   .193011
3 H   .193011
Sum of Mulliken charges = .00000
Atomic charges with hydrogens summed into heavy atoms:
    1
1 O   .000000
2 H   .000000
3 H   .000000
Sum of Mulliken charges = .00000
Electronic spatial extent (au): $<R**2> = 17.7491$
Charge = .0000 electrons
Dipole moment (Debye):
X =   .0000   Y =   .0000   Z =   -1.7634   Tot =   1.7634
Quadrupole moment (Debye-Ang):
XX =   -6.0801   YY =   -4.3053   ZZ =   -5.3664
XY =   .0000   XZ =   .0000   YZ =   .0000
Octapole moment (Debye-Ang**2):
XXX =   .0000   YYY =   .0000   ZZZ =   -.2251   XYY =   .0000
XXY =   .0000   XXZ =   -.0089   XZZ =   .0000   YZZ =   .0000
YYZ =   -.5756   XYZ =   .0000
Hexadecapole moment (Debye-Ang**3):
XXXX =   -3.2323   YYYY =   -6.4280   ZZZZ =   -4.8602   XXXY =   .0000
XXXZ =   .0000   YYYX =   .0000   YYYZ =   .0000   ZZZX =   .0000
ZZZY =   .0000   XXYY =   -1.7655   XXZZ =   -1.3841   YYZZ =   -1.6675
XXYZ =   .0000   YYXZ =   .0000   ZZXY =   .0000
N-N = 9.206554604692D + 00   E-N = -1.985365895863D + 02   KE = 7.549085547930D + 01
Symmetry A1 KE = 6.746608229391D + 01
Symmetry A2 KE = .000000000000D + 00
Symmetry B1 KE = 5.063673585135D + 00
Symmetry B2 KE = 2.961099600255D + 00
1\1\GINC-UCHHPA\SP\RHF\STO-4G\H2O1\MCDTSAH\12–Jul-1994\0\\# HF/STO-4G
POP = FULL\\WATER MOLECULE\\0,1\O\H,1,0.956\H,1,0.956,2,104.5\\Version = H
P-PARisc-HPUX-G92RevE.1\State = 1–A1\HF = -75.4964937\RMSD = 7.012e-04\Dipole
= 0.5485633,0.,0.4247431\PG = C02V [C2(O1),SGV(H2)]\\

Job cpu time: 0 days 0 hours 0 minutes 5.4 seconds.
File lengths (MBytes): RWF = 5   Int = 0   D2E = 0   Chk = 1   Scr = 0 | Timing |
Normal termination of Gaussian 92.

9.2.1 Explanation of output

I have annotated the output with boxes to explain what is happening.

Route

GAUSSIAN92 is a very sophisticated package, and it is possible to pick and match various calculations (a 'route') within the general framework of molecular structure theory. The first record # HF/STO–4G . . . of our very simple job defines our preset route through the package. May I remind you of the two-electron integrals problem discussed earlier. There are usually an awful lot of them, despite the fact that they are easy to calculate with Gaussian orbitals. In this example, the basis set comprises seven basis functions made up from 28 primitive Gaussian orbitals. The maximum number of two-electron integrals over the basis functions in this simple case is only 4060, but many of these can be deduced by symmetry to be zero. These integrals can be stored on disk, stored in memory or recalculated each cycle of the SCF procedure. In this example, I arranged for the integrals to be recalculated each cycle of the SCF procedure (SCF = Direct).

Molecular geometry

The package has to translate the Z-matrix into cartesian coordinates. It then calculates bond distances, and it is well worth spending a little time reading this section of the output just in case of errors. Plot out the atoms yourself, the package won't do it for you. Does everything appear to be in order? Up to this point no use has been made of molecular symmetry.

Molecular symmetry

The next step is to try and make use of molecular symmetry. H_2O in this calculation is assumed to have C_{2v} symmetry, because I have chosen equal OH bond distances. The package makes use of molecular symmetry at various points in the calculation, but this is usually transparent to users unless they are specifically looking for it. See my comments above regarding the loss of symmetry due to your Z-matrix.

How do you know that H_2O should really have C_{2v} symmetry? Why did I start off the calculation with both O-H bonds equal? If you want to investigate such questions, you will have to be careful in the way you write your Z-matrix. The way I have written the Z-matrix, the two O-H bonds will certainly be treated as equal.

SCF cycles

The program default is to use a guess at the electron density calculated using the INDO model discussed in Chapter 7. The program iterates until the electron density is self consistent, and then outputs the results shown. The kinetic energy in this case is 75.491 E_h, the one-electron potential energy is $-198.535E_h$ and the 'total' energy, including the nuclear repulsion is $-75.496E_h$.

LCAO-MO coefficients

LCAO-MO coefficients are then output. Notice that the package has taken all the hard work out of the group theory and labelled each MO according to its irreducible representation in the C_{2v} point group. The first LCAO-MO is the first column shown, and is 0.99592 oxygen 1s, 0.01920 oxygen 2s etc. The –O means an occupied orbital, and the –V means an empty ('virtual') orbital. The virtual orbitals are not generally useful.

In the spirit of elementary valence theory, The molecular orbital configuration diagram is shown in Figure 9.3. There are five occupied orbitals and I have shown the first virtual one.

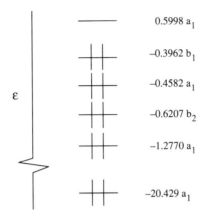

Figure 9.3   The molecular orbital configuration

Charge density

Finally, we need to make some attempt to understand the electron density. You should be able to understand this section in terms of our discussion of earlier chapters. Of particular interest are the Mulliken 'total atomic charges' of -0.386022 (oxygen) and $+0.193011$ (per hydrogen). If you wish to divide up the Mulliken electron populations into contributions from the individual orbitals, the data is given as 'gross orbital populations'. The atomic charges are found by subtracting the orbital contributions from the formal atomic charge.

The package then calculates various electronic properties such as the electric dipole moment. I will discuss these properties in Chapter 13.

Timing

Timing is obviously important, especially if you are paying for the calculation with real money!. In this case, the calculation took 5.4 seconds on a Hewlett-Packard Apollo 700 workstation.

9.2.2 Resource usage

One of the messages of this text is the correct choice of basis set for a molecular calculation. Roughly speaking, the larger the basis set the more 'accurate' will be the prediction, other things being equal. Care is needed; if the molecular property under study depends critically on the outer regions of the electron density, there is probably little point in using an energy-optimized minimal basis set. This is because energy-optimized basis sets are usually constructed using the variation principle. The process optimizes wavefunctions according to the electronic Hamiltonian, and the 'nuclear–

electron' attraction terms tend to dominate in order to give the best energy. The variational optimization process normally focuses attention on the inner regions of the electron density.

The cost of a calculation depends on a number of things, and if we keep the level of theory constant at the HF model, you might be interested to see how the cost (and the total energy) vary with basis set size and sophistication for a larger molecule, the all-*trans* $C_{10}H_{12}$ polyene (Table 9.1).

Table 9.1 $C_{10}H_{12}$ deca-1,3,5,7,9-ene all-trans SCF GAUSSIAN92 calculations

Basis set	Basis functions	Primitives	Energy/E_h	time/s
STO/3G	62	186	−380.860392	32
STO/4G	62	248	−383.614088	53
STO/6G	62	372	−384.602975	99
STO/4–31G	114	248	−385.036732	117
STO/6–31G	114	268	−385.447757	135
D95V	114	298	−385.477206	322
STO/6–31G*	174	328	−385.585782	462
6–311G**	252	406	−385.678071	1714

The presence of polarization functions in a heavy-atom basis set is denoted by a * and in a hydrogen atom by a further *.

Various trends are apparent. As we improve the inner shells by moving across the series STO–nG, the total energy decreases rapidly. The split valence calculations usually give a lower energy than the STO/nG ones and so on. The resource consumption of a HF calculation is usually quoted as being proportional to $n^{3.5}$, where n is the number of basis functions. The reason for this statement is that it is necessary to calculate about that number of two-electron integrals for the HF calculation. The dominant part of a HF calculation is usually that taken by integral evaluation.

9.3   THE HARTREE FOCK LIMIT

We can rank the calculations above in order of 'goodness' according to the variation principle; the lower the energy the 'better'. Thus we note, for example, the dramatic improvement in moving along the series STO/nG. This is caused by improving the description of the carbon inner shell electron density. Eventually the gains in total energy become smaller with increase in sophistication of basis set and with cost.

The experimental value of the total energy is unknown and, in any case, direct comparisons between calculated values and the experimental value would have to be made with some caution. The HF calculations refer to a fixed geometry and a non-relativistic Hamiltonian. In HF calculations the electrons are taken to move in an average potential due to the remaining electrons. The best HF calculation that could possibly be made (i.e., the 'HF limit') would still give an energy higher than the true one. The HF limit can be reached readily in **atomic** SCF calculations where the Hartree

Fock equations can be integrated numerically. For molecules, however, calculations at the Hartree Fock limit are unattainable.

The difference between the 'experimental' and the Hartree Fock limit energies is called the **correlation energy** and it is typically a few per cent of the total energy. Perhaps the most important consequence is that SCF calculations **cannot**, in principle, treat phenomena such as dispersion forces, which depend on the **instantaneous** interactions between particles.

9.4 OPEN SHELL CALCULATIONS

The lowest energy cation of H_2O is formed by removing an electron from the $1b_1$ molecular orbital (in MO language). If we assume for now that the cation geometry is identical to that of the parent molecule, all that is necessary to perform an open shell calculation is to specify ROHF (restricted open shell Hartree Fock) or UHF (unrestricted Hartree Fock) instead of HF in the # record.

9.4.1 A ROHF GAUSSIAN92 run

An STO/4G ROHF calculation gives pretty much the same output as before (I have condensed it a little) except that we now get the spin densities output in addition to the charge densities. I will return to the theme of spin density in Chapter 15.

```
* * * * * * * * * * * * * * * * * * * * * * * * * * * * * * * * * * * * * * * * * *
Gaussian 92: HP-PARisc-HPUX-G92RevE.1 12–Jun-1993
            12–Jul-1994
* * * * * * * * * * * * * * * * * * * * * * * * * * * * * * * * * * * * * * * * * *
%save
Default route: SCF = Direct MP2 = Stingy MAXDISK = 25000000
-----------------
# ROHF/STO-4G POP = FULL
-----------------
1/29 = 10000/1;
2/12 = 2,17 = 6,18 = 5/2;
3/6 = 4,11 = 2,25 = 1,30 = 1/1,2,3;
4/7 = 6/1;
5/5 = 2,32 = 1,38 = 4/2;
6/7 = 3,19 = 1,28 = 1/1;
99/5 = 1,9 = 1/99;
-----------------------------------------------------------------
WATER MOLECULE lowest energy cation
-----------------------------------------------------------------
Symbolic Z-matrix:
Charge = 1 Multiplicity = 2
O
H1  O  ROH
H2  O  ROH  H1  ANG
   Variables:
ROH          0.956
ANG          104.5
```

. . . .

A Direct SCF calculation will be performed.
Using DIIS extrapolation.
Restricted open shell SCF:
Requested convergence on RMS density matrix = 1.00D-04 within
64 cycles.
Requested convergence on MAX density matrix = 1.00D-02.
Requested convergence on   energy = 5.00D-05.
Two-electron integral symmetry used by symmetrizing Fock
matrices.
Keep R1 and R2 integrals in memory in canonical form, NReq = 430344.
IEnd = 28546 IEndB = 28546 NGot = 2048000 MDV = 2027143
LenX = 2027143
MinBra = 0 MaxBra = 3 MinLOS = -1 MaxLOS = -1 MinRaf = 0 MaxRaf = 3 MinLRy = 4.
IRaf = 0 NMat = 1 IRICut = 1 DoRegI = T DoRafI = F ISym2E = 1 JSym2E = 1.
Fock matrices symmetrized in FoFDir.
Convergence on energy, delta-E = 2.10D-05
SCF DONE: E(ROHF) = -75.1002855786 A.U. AFTER 5 CYCLES
CONVG = .3835D-03 -V/T = 2.0147
S**2 = .7500
KE = 7.401245767058D + 01 PE = -1.912956564042D + 02 EE = 3.297635855031D + 01
Annihilation of the first spin contaminant:
S**2 BEFORE ANNIHILATION .7500, AFTER .7500
Copying SCF densities to generalized density rwf, ISCF = 0 IROHF = 1.

* *

Population analysis using the SCF density.

* *

ORBITAL SYMMETRIES.
   OCCUPIED (A1) (A1) (B2) (B1) (A1)
   VIRTUAL (A1) (B2)
THE ELECTRONIC STATE IS 2-A1.
Alpha occ. eigenvalues — -21.14607 -1.78469 -1.14945 -.95724 -.49695
Alpha virt. eigenvalues — .12411 .21510
Molecular Orbital Coefficients

			1	2	3	4	5
			(A1)–O	(A1)–O	(B2)–O	(B1)–O	(A1)–O
		EIGENVALUES —	21.14607	-1.78469	-1.14945	-.95724	-.49695
1 1	O 1S		.99617	-.22604	.00000	.00000	-.09111
2	2S		.01800	.86446	.00000	.00000	.45962
3	2PX		.00000	.00000	.00000	1.00000	.00000
4	2PY		.00000	.00000	.72057	.00000	.00000
5	2PZ		-.00264	-.19600	.00000	.00000	.85367
6 2	H 1S		-.00400	.12121	.34172	.00000	-.17567
7 3	H 1S		-.00400	.12121	-.34172	.00000	-.17567

			6	7
			(A1)–V	(B2)–V
		EIGENVALUES –	.12411	.21510
1 1	O 1S		-.12235	.00000
2	2S		.90095	.00000
3	2PX		.00000	.00000
4	2PY		.00000	.91072
5	2PZ		-.63607	.00000
6 2	H 1S		-.83403	-.88611
7 3	H 1S		-.83403	.88611

LCAO-MO

ALPHA DENSITY MATRIX.

			1	2	3	4	5
1 1	O	1S	1.05175				
2		2S	-.21935	.95886			
3		2PX	.00000	.00000	1.00000		
4		2PY	.00000	.00000	.00000	.51922	
5		2PZ	-.03610	.22288	.00000	.00000	.76717
6 2	H	1S	-.01537	.02397	.00000	.24624	-.17371
7 3	H	1S	-.01537	.02397	.00000	-.24624	-.17371

Charge density

			6	7
6 2	H	1S	.16234	
7 3	H	1S	-.07121	.16234

BETA DENSITY MATRIX.

			1	2	3	4	5
1 1	O	1S	1.04345				
2		2S	-.17747	.74761			
3		2PX	.00000	.00000	1.00000		
4		2PY	.00000	.00000	.00000	.51922	
5		2PZ	.04168	-.16948	.00000	.00000	.03842
6 2	H	1S	-.03138	.10471	.00000	.24624	-.02375
7 3	H	1S	-.03138	.10471	.00000	-.24624	-.02375

			6	7
6 2	H	1S	.13148	
7 3	H	1S	-.10207	.13148

Full Mulliken population analysis:

			1	2	3	4	5
1 1	O	1S	2.09520				
2		2S	-.09183	1.70647			
3		2PX	.00000	.00000	2.00000		
4		2PY	.00000	.00000	.00000	1.03845	
5		2PZ	.00000	.00000	.00000	.00000	.80559
6 2	H	1S	-.00258	.06141	.00000	.15309	.04753
7 3	H	1S	-.00258	.06141	.00000	.15309	.04753

			6	7
6 2	H	1S	.29382	
7 3	H	1S	-.04373	.29382

Gross orbital populations:

			TOTAL	ALPHA	BETA	SPIN
1 1	O	1S	1.99822	.99930	.99892	.00037
2		2S	1.73745	.93097	.80647	.12450
3		2PX	2.00000	1.00000	1.00000	.00000
4		2PY	1.34462	.67231	.67231	.00000
5		2PZ	.90065	.85079	.04986	.80094
6 2	H	1S	.50953	.27331	.23622	.03709
7 3	H	1S	.50953	.27331	.23622	.03709

Condensed to atoms (all electrons):

	1	2	3
1 O	7.462049	.259442	.259442
2 H	.259442	.293824	-.043734
3 H	.259442	-.043734	.293824

Total atomic charges:

	1
1 O	.019066

2 H   .490467
3 H   .490467
Sum of Mulliken charges = 1.00000
Atomic charges with hydrogens summed into heavy atoms:
          1
1 O   1.000000
2 H   .000000
3 H   .000000
Sum of Mulliken charges = 1.00000
Atomic-Atomic Spin Densities.
         1   2   3
1 O   .928919 -.001554 -.001554
2 H   -.001554 .030860 .007789
3 H   -.001554 .007789 .030860

Spin densities

Total atomic spin densities:
         1
1 O   .925810
2 H   .037095
3 H   .037095
Sum of Mulliken spin densities = 1.00000
Fermi contact analysis (atomic units).
         1
1 O   .778210
2 H   .025039
3 H   .025039
Electronic spatial extent (au): $<R**2> = 13.9704$
Charge = 1.0000 electrons
Dipole moment (Debye):
X = .0000 Y = .0000 Z = -1.9341 Tot = 1.9341
Quadrupole moment (Debye-Ang):
XX = -5.3997 YY = -2.1357 ZZ = -3.1338
XY = .0000   XZ = .0000 YZ = .0000
Octapole moment (Debye-Ang**2):
XXX = .0000   YYY = .0000 ZZZ = -.7676 XYY = .0000
XXY = .0000   XXZ = -.1450 XZZ = .0000   YZZ = .0000
YYZ = -1.5877 XYZ = .0000
Hexadecapole moment (Debye-Ang**3):
XXXX = -2.7812 YYYY = -3.3410 ZZZZ = -2.8666 XXXY = .0000
XXXZ = .0000   YYYX = .0000   YYYZ = .0000   ZZZX = .0000
ZZZY = .0000   XXYY = -1.3358 XXZZ = -1.0082 YYZZ = -.6293
XXYZ = .0000   YYXZ = .0000   ZZXY = .0000
Job cpu time: 0 days 0 hours 0 minutes 5.2 seconds.
File lengths (MBytes): RWF = 5 Int = 0 D2E = 0 Chk = 1 Scr = 0
Normal termination of Gaussian 92.

For the minute, note that the total energy is $-75.100286E_h$. The variation method produces the lowest energy state of a given symmetry, by which I mean **spatial** (in this case the lowest energy $^2A_1$ electronic state) or **spin** (so I could have also studied the lowest energy $^4A_1$ state). GAUSSIAN92 outputs the electron density matrices for the alpha and beta spin electrons separately and reports a full Mulliken population analysis as before.

Of particular interest is **the spin density**, mentioned in an earlier chapter. In electron spin resonance experiments, provided that the nucleus of interest has a non-zero

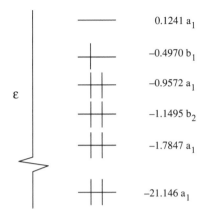

Figure 9.4 Orbital ebergy diagram for H_2O^+, calculated at the ROHF level of theory

nuclear magnetic dipole, the isotropic coupling constant depends on the value of the spin density function evaluated at the position of that particular nucleus. These values are printed.

9.4.2 An UHF GAUSSIAN92 run

Finally, let's switch the route to UHF rather than ROHF. You should remember the aim of UHF theory; the α and β spin electrons have different spatial parts to their descriptions. The only parts of the output that are significantly different from the ROHF case are shown below

---

UHF/STO-4G POP = FULL

WATER MOLECULE lowest energy cation

Symbolic Z-matrix:
Charge = 1 Multiplicity = 2
Two-electron integral symmetry is turned on.
   7 basis functions   28 primitive gaussians
   5 alpha electrons   4 beta electrons
     nuclear repulsion energy   9.2065546047 Hartrees.
One-electron integrals computed using PRISM.
One-electron integral symmetry used in STVInt
The smallest eigenvalue of the overlap matrix is 3.412D-01
DipDrv: MaxL = 4.
DipDrv: will hold 34 matrices at once.
PROJECTED INDO GUESS.
INITIAL GUESS ORBITAL SYMMETRIES.
  ALPHA ORBITALS
    OCCUPIED (A1) (A1) (B2) (B1) (A1)
    VIRTUAL (A1) (B2)
  BETA ORBITALS
    OCCUPIED (A1) (A1) (B2) (A1)
    VIRTUAL (B1) (A1) (B2)

< S**2 > of initial guess = .7535
Alpha deviation from unit magnitude is 4.44D-16 for orbital 6.
Alpha deviation from orthogonality is 2.22D-16 for orbitals 7 2.
Beta deviation from unit magnitude is 2.22D-16 for orbital 5.
Beta deviation from orthogonality is 1.11D-16 for orbitals 3 2.
Warning! Cutoffs for single-point calculations used.
A Direct SCF calculation will be performed.
Using DIIS extrapolation.
UHF open shell SCF:
Requested convergence on RMS density matrix = 1.00D-04 within 64 cycles.
Requested convergence on MAX density matrix = 1.00D-02.
Requested convergence on energy = 5.00D-05.
Two-electron integral symmetry used by symmetrizing Fock matrices.
Keep R1 and R2 integrals in memory in canonical form, NReq = 430344.
IEnd = 28546 IEndB = 28546 NGot = 2048000 MDV = 2027143
LenX = 2027143
MinBra = 0 MaxBra = 3 MinLOS = -1 MaxLOS = -1 MinRaf = 0 MaxRaf = 3 MinLRy = 4.
IRaf = 0 NMat = 1 IRICut = 1 DoRegI = T DoRafI = F ISym2E = 1 JSym2E = 1.
Fock matrices symmetrized in FoFDir.
Convergence on energy, delta-E = 3.75D-05
SCF DONE: E(UHF) = -75.1846410426 A.U. AFTER 4 CYCLES
CONVG = .7248D-03 -V/T = 2.0165
S**2 = .7550
KE = 7.396446180439D + 01 PE = -1.910567260755D + 02 EE = 3.270106862385D + 01
Annihilation of the first spin contaminant:
S**2 BEFORE ANNIHILATION .7550, AFTER .7500
Copying SCF densities to generalized density rwf, ISCF = 1 IROHF = 0.
* *
Population analysis using the SCF density.
* *
ORBITAL SYMMETRIES.
  ALPHA ORBITALS
    OCCUPIED (A1) (A1) (B2) (B1) (A1)
    VIRTUAL (A1) (B2)
  BETA ORBITALS
    OCCUPIED (A1) (A1) (B2) (A1)
    VIRTUAL (B1) (A1) (B2)
THE ELECTRONIC STATE IS 2–B1.
Alpha occ. eigenvalues — 21.22054   -1.93830   -1.20456   -1.11999   -1.07360
Alpha virt. eigenvalues –   .07907   .19166
Beta occ. eigenvalues –   -21.18871   -1.76764   -1.16235   -1.01452
Beta virt. eigenvalues –   -.23153   .11137   .21193
Alpha Molecular Orbital Coefficients

		1	2	3	4	5
		(A1)–O	(A1)–O	(B2)–O	(B1)–O	(A1)–O
	EIGENVALUES —	21.22054	-1.93830	-1.20456	-1.11999	-1.07360
1 1	O 1S	.99587	-.23581	.00000	.00000	-.07947
2	2S	.01943	.91926	.00000	.00000	.41840
3	2PX	.00000	.00000	.00000	1.00000	.00000
4	2PY	.00000	.00000	.74476	.00000	.00000
5	2PZ	-.00349	-.17375	.00000	.00000	.85929
6 2	H 1S	-.00443	.08170	.31781	.00000	-.18641
7 3	H 1S	-.00443	.08170	-.31781	.00000	-.18641

	6	7
	(A1)–V	(B2)–V

```
      EIGENVALUES –  .07907  .19166
1 1  O 1S              -.11420    .00000
2     2S                .86608    .00000
3     2PX               .00000    .00000
4     2PY               .00000   -.89104
5     2PZ              -.63496    .00000
6 2  H 1S             -.83650    .89497
7 3  H 1S             -.83650   -.89497
```
Beta Molecular Orbital Coefficients.

		1	2	3	4	5
		(A1)–O	(A1)–O	(B2)–O	(A1)–O	(B1)–V
	EIGENVALUES —	21.18871	-1.76764	-1.16235	-1.01452	-.23153
1 1	O 1S	.99643	-.22381	.00000	-.09158	.00000
2	2S	.01687	.85264	.00000	.47342	.00000
3	2PX	.00000	.00000	.00000	.00000	1.00000
4	2PY	.00000	.00000	.70202	.00000	.00000
5	2PZ	-.00316	-.18141	.00000	.84159	.00000
6 2	H 1S	-.00386	.13337	.35956	-.19313	.00000
7 3	H 1S	-.00386	.13337	-.35956	-.19313	.00000

		6	7
		(A1)–V	(B2)–V
	EIGENVALUES –	.11137	.21193
1 1	O 1S	-.12396	.00000
2	2S	.90508	.00000
3	2PX	.00000	.00000
4	2PY	.00000	.92509
5	2PZ	-.65617	.00000
6 2	H 1S	-.82830	-.87903
7 3	H 1S	-.82830	.87903

ALPHA DENSITY MATRIX.

		1	2	3	4	5
1 1	O 1S	1.05368				
2	2S	-.23067	1.02048			
3	2PX	.00000	.00000	1.00000		
4	2PY	.00000	.00000	.00000	.55467	
5	2PZ	-.03079	.19974	.00000	.00000	.76858
6 2	H 1S	-.00886	-.00298	.00000	.23669	-.17436
7 3	H 1S	-.00886	-.00298	.00000	-.23669	-.17436

		6	7
6 2	H 1S	.14244	
7 3	H 1S	-.05956	.14244

BETA DENSITY MATRIX.

		1	2	3	4	5
1 1	O 1S	1.05135				
2	2S	-.21738	.95140			
3	2PX	.00000	.00000	.00000		
4	2PY	.00000	.00000	.00000	.49284	
5	2PZ	-.03962	.24370	.00000	.00000	.74119
6 2	H 1S	-.01601	.02222	.00000	.25242	-.18672
7 3	H 1S	-.01601	.02222	.00000	-.25242	-.18672

		6	7
6 2	H 1S	.18438	
7 3	H 1S	-.07418	.18438

Full Mulliken population analysis:

		1	2	3	4	5
1 1	O 1S	2.10503				

2	2S	-.10368	1.97188			
3	2PX	.00000	.00000	1.00000		
4	2PY	.00000	.00000	.00000	1.04751	
5	2PZ	.00000	.00000	.00000	.00000	1.50977
6 2	H 1S	-.00137	.00918	.00000	.15204	.08691
7 3	H 1S	-.00137	.00918	.00000	.15204	.08691

		6	7
6 2	H 1S	.32683	
7 3	H 1S	-.03376	.32683

Gross orbital populations:

		TOTAL	ALPHA	BETA	SPIN
1 1	O 1S	1.99860	.99932	.99928	.00004
2	2S	1.88656	.96425	.92231	.04195
3	2PX	1.00000	1.00000	.00000	1.00000
4	2PY	1.35160	.70183	.64977	.05206
5	2PZ	1.68358	.85251	.83107	.02144
6 2	H 1S	.53983	.24105	.29878	-.05774
7 3	H 1S	.53983	.24105	.29878	-.05774

Spin densities

Condensed to atoms (all electrons):

		1	2	3
1	O	7.426819	.246761	.246761
2	H	.246761	.326825	-.033756
3	H	.246761	-.033756	.326825

Total atomic charges:

		1
1	O	.079660
2	H	.460170
3	H	.460170

Sum of Mulliken charges = 1.00000
Atomic charges with hydrogens summed into heavy atoms:

		1
1	O	1.000000
2	H	.000000
3	H	.000000

Sum of Mulliken charges = 1.00000
Atomic-Atomic Spin Densities.

		1	2	3
1	O	1.154465	-.019493	-.019493
2	H	-.019493	-.041936	.003690
3	H	-.019493	.003690	-.041936

Total atomic spin densities:

		1
1	O	1.115479
2	H	-.057739
3	H	-.057739

Sum of Mulliken spin densities = 1.00000
Fermi contact analysis (atomic units).

		1
1	O	.219946
2	H	-.023467
3	H	-.023467

Electronic spatial extent (au): $< R**2 > = 14.3943$
Charge = 1.0000 electrons
Dipole moment (Debye):
X = .0000 Y = .0000 Z = -2.4991 Tot = 2.4991
Quadrupole moment (Debye-Ang):

XX = -4.6652 YY = -2.3640 ZZ = -4.2101
XY = .0000 XZ = .0000 YZ = .0000
Octapole moment (Debye-Ang**2):
XXX = .0000 YYY = .0000 ZZZ = -1.0732 XYY = .0000
XXY = .0000 XXZ = -.0925 XZZ = .0000 YZZ = .0000
YYZ = -1.5054 XYZ = .0000
Hexadecapole moment (Debye-Ang**3):
XXXX = -2.3044 YYYY = -3.6585 ZZZZ = -3.7654 XXXY = .0000
XXXZ = .0000   YYYX = .0000   YYYZ = .0000   ZZZX = .0000
ZZZY = .0000   XXYY = -1.2925 XXZZ = -1.0542 YYZZ = -.8566
XXYZ = .0000   YYXZ = .0000   ZZXY = .0000
N-N = 9.206554604692D + 00 E-N = -1.910568365239D + 02 KE = 7.396446180439D + 01
Job cpu time: 0 days 0 hours 0 minutes 5.6 seconds.
File lengths (MBytes): RWF = 5 Int = 0 D2E = 0 Chk = 1 Scr = 0
Normal termination of Gaussian 92.

Note the total energy of $-75.18464E_h$, and the fact that the α and β spin electrons are truly treated differently; each set of electrons has its own orbital diagram like the one above. Again, notice the Mulliken population analysis, and the spin density analysis. But there is a new message concerning the expectation value $<S^2>$. HF and ROHF wavefunctions are eigenfunctions of both the spin operator S_z and S^2. UHF wavefunctions are only eigenfunctions of S_z, and the expectation value $<S^2>$ tells us how much the UHF process has produced a mixture of different spin states. The UHF calculation on H_2O is trying to describe a spin doublet state. The UHF procedure gives a mixture

$$\Psi(\text{UHF}) = c_2\Psi_2(\text{doublet}) + c_4\Psi_4(\text{quartet}) + c_6\Psi_6(\text{hextet}) + \ldots$$

and the pious hope is that the doublet state will dominate. In the case of our H_2O calculation, this seems to be the case because the $<S^2>$ value comes out as 0.7550 compared with the 'theoretical' value of $\frac{1}{2}(\frac{1}{2}+1) = \frac{3}{4}$ (Table 9.2).

Table 9.2 H_2O STO/4G calculations

State	Method	E (E_h)	$<S^2>$
$^1A_1$	HF	−75.496494	$\frac{3}{4}$
$^2B_1$	ROHF	−75.100286	$\frac{3}{4}$
$^2B_1$	UHF	−75.184641	0.7550

9.5 A GUIDE TO THE LITERATURE

The *Quantum Chemistry Literature Data Base (QCLDB)* is published by Elsevier as a series of supplements to the original book *QCLDB—Bibliography of* Ab Initio *Calculations for 1978–1980*. The most recent is Supplement 13, which is a bibliography of calculations for 1993. Single copies can be purchased from the publisher.

10 Electron Correlation

In our discussion of the electron density in Chapter 4, I mentioned the density functions $\rho_1(\underline{x}_1)$ and $\rho_2(\underline{x}_1, \underline{x}_2)$. These have a probabilistic interpretation; $\rho_1(\underline{x}_1)d\tau_1 ds_1$ gives the chance of finding an electron in the element $d\tau_1 ds$ of space and spin, whilst $\rho_2(\underline{x}_1, \underline{x}_2)d\tau_1 ds_1$ determines the probability of two electrons being found simultaneously in the elements $d\tau_1 ds_1$ and $d\tau_2 ds_2$. The two-electron function gives information as to how the motions of any pair of electrons are 'correlated' as a result of their electrostatic interactions. For independent particles,

$$\rho_2(\underline{x}_1, \underline{x}_2) = \rho_1(\underline{x}_1)\rho_1(\underline{x}_2)$$

and we say in such a case that there is **no correlation** between the electrons. This result holds for electrons in the original Hartree SCF model, where the effects of antisymmetry are ignored. In the Hartree Fock method, it turns out that the motions of electrons of **like** spin are correlated and this prevents the two electrons of like spin being found at the same point in space. On the other hand there is no correlation between the spatial positions of electrons of opposite spins, which is obviously a defect. This type of correlation is often referred to as **Fermi correlation**.

It is clear that the HF energy will always be greater than the true energy, no matter how sophisticated the HF methodology because of the error inherent in HF calculations. We call this (positive) energy difference the **correlation energy**. So, HF theory does not treat electron correlation properly. The remarkable thing is that HF calculations are so reliable for the calculation of many molecular properties, as I will discuss in later chapters.

But for many simple applications, a more advanced treatment of electron correlation is essential. In Chapter 3, we dealt with the simple LCAO treatment of H_2 and calculated the potential energy curve reproduced in Figure 10.1. The lowest energy 'products' for the dissociation of dihydrogen are atoms in their lowest electronic state.

$$H_2 \rightarrow 2H(^2s)$$

In the atomic system of units, the energy of the products is $2 \times (-\frac{1}{2})$ E_h, so the potential curve should tend asymptotically to $-1E_h$. An analysis of the results shows that the HF wavefunction is dissociating into ions

$$H_2 \rightarrow H^+ + H^-$$

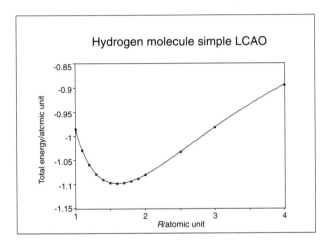

Figure 10.1 Simple H_2 LCAO treatment, (from Chapter 3)

This turns out to be common behaviour; whenever strong bonds are made or broken, the HF wavefunction will tend to give incorrect dissociation products. I showed you how to correct the problem using configuration interaction. We write suitable wavefunctions corresponding to $1\sigma_g^2$, $1\sigma_g^1 1\sigma_u^1$ (a **singly excited state**) and $1\sigma_u^2$ (a **doubly excited state**) and mix them using the variation method. Because of the symmetry of the problem, only the doubly excited state mixed in with the ground state.

As a second simple example, consider the ground state of C_2, which we might expect to be

C_2; $1\sigma_g^2 1\sigma_u^2 2\sigma_g^2 2\sigma_u^2 1\pi_u^4$ $^1\Sigma_g^+$

But two other nearby configurations are also possible contenders

C_2; $1\sigma_g^2 1\sigma_u^2 2\sigma_g^2 2\sigma_u^2 1\pi_u^3 3\sigma_g^1$ $^1\Pi_u$, $^3\Pi_u$

C_2 $1\sigma_g^2 1\sigma_u^2 2\sigma_g^2 2\sigma_u^2 1\pi_u^2 3\sigma_g^2$ $^3\Sigma_g^-$, $^1\Delta_g$, $^1\Sigma_g^+$

Even at the equilibrium geometry, we will have to treat the two $^1\Sigma_g^+$ states in order to get a reasonable description of the electronic wavefunction.

As a final example, think back to the electronic excited states of a simple π-electron molecule (pyridine) in Chapter 7. The aim was then to study the ground state, the excited states and properties such as the transition energies. In this particular example, we wanted to know details of all these states, not just the ground one. I did this calculation within the ZDO approximation.

The steps we followed in our HF-CI treatment of H_2 were these

- choose a molecular geometry;
- choose a basis set (in this case, simple STOs on either centre);
- calculate the HF orbitals (determined by symmetry in this particular case);
- choose excited configurations;
- calculate the relevant Hamiltonian matrix;
- find its eigenvalues and eigenvectors.

In the fourth step, I just selected singly and doubly excited configurations, $1\sigma_g^2$ for the ground state electronic configuration, $1\sigma_g^1 1\sigma_u^1$ for a singly excited state and $1\sigma_u^2$ for the first available doubly excited state. I then sought to improve the ground state wavefunction by writing a 'better' electronic wavefunction as

$$\Psi_{\text{better}} = a\Psi(1\sigma_g^2) + b\Psi(1\sigma_g^1 1\sigma_u^1) + c\Psi(1\sigma_u^2)$$

where a,b and c have to be determined from the variation principle.

In the fifth step, I had to work out a 3×3 Hamiltonian matrix with elements typically

$$\int \Psi(1\sigma_g^2) H \psi(1\sigma_u^2) d\tau$$

and so we need to express the excited states and the ground state as Slater determinants, and then evaluate them using the integrals calculated over the basis functions. Finally, we diagonalize the matrix to recover information about the electronic state(s) of interest. Perhaps we are only interested in the ground state, perhaps we want to find the energy of the various excited states relative to the ground state.

It is the fifth step that leads to the problem for larger molecules. The reason is easy to grasp, as I discussed in our ZDO treatment of pyridine in Chapter 7. The energy formulae will involve both one-electron integrals and two electron integrals over the HF LCAO orbitals $\psi_A \psi_B \ldots \psi_M$. I showed that a matrix element between two singly excited states formed by exciting an electron from ψ_A to ψ_X and from ψ_B to ψ_Y gave

$$\int \Psi(A \to X) H \psi(B \to Y) d\tau = -\int\int \psi_A(\mathbf{r}_1)\psi_B(\mathbf{r}_1) g(\mathbf{r}_1,\mathbf{r}_2)\psi_X(\mathbf{r}_2)\psi_Y(\mathbf{r}_2) d\tau_1 d\tau_2$$
$$+ 2\int\int \psi_A(\mathbf{r}_1)\psi_X(\mathbf{r}_1) g(\mathbf{r}_1,\mathbf{r}_2)\psi_B(\mathbf{r}_2)\psi_Y(\mathbf{r}_2) d\tau_1 d\tau_2$$

The major time-consuming part of a conventional CI calculation is the transformation of the electron repulsion integrals calculated over the basis functions to integrals calculated over the LCAO-MOs. So, for example, a typical two-electron integral

$$\int\int \psi_A(\mathbf{r}_1)\psi_B(\mathbf{r}_1) g(\mathbf{r}_1,\mathbf{r}_2)\psi_X(\mathbf{r}_2)\psi_Y(\mathbf{r}_2) d\tau_1 d\tau_2$$

would have to be evaluated by somehow performing the sum over basis functions χ according to

$$\sum \sum \sum \sum a_i b_j x_k y_l \int \int \chi_i(\mathbf{r}_1)\chi_j(\mathbf{r}_1)g(\mathbf{r}_1,\mathbf{r}_2)\chi_k(\mathbf{r}_2)\chi_l(\mathbf{r}_2)\mathrm{d}\tau_1\mathrm{d}\tau_2$$

where the values of a, b, c and d are LCAO coefficients. This sum appears at first sight to contain n^4 terms. Many of the popular methods for treating correlation in larger molecules are those that can successfully negotiate this **integral transformation problem**. I will describe just two such methods to try and give the flavour.

10.1 THE MØLLER PLESSET METHOD

The Møller Plesset method is probably the most popular way of addressing electron correlation in real molecules, and you might like to read the original paper.

Note on an Approximation Treatment for Many-Electron Systems
Chr Møller and M. S. Plesset
Physical Review, **46** (1934) 618

A Perturbation theory is developed for treating a system of n electrons in which the Hartree Fock solution appears as the zero-order approximation. It is shown by this development that the first order correction for the energy and the charge density of the system is zero. The expression for the second order correction for the energy greatly simplifies because of the special property of the zero order solution. It is pointed out that the development of the higher order approximation involves only calculations based on a definite one-body problem.

It is couched in the language of perturbation theory, so I had better remind you of the key concepts.

10.1.1 Perturbation theory

The number of problems which can be solved exactly by quantum mechanical methods is not very large. Suppose that our problem is to solve

$$H\Psi_i = \epsilon_i \Psi_i$$

which does not seem to have an easy solution, whilst a simpler related problem

$$H^0 \Psi_i^0 = \epsilon_i^0 \Psi_i^0$$

can be solved exactly. For example, H might be the Hamiltonian for a He atom, and the **zero order problem** with the superscript 0 might refer to two superimposed H atoms whose electrons did not interact. The idea is to write the Hamiltonian H as

$$H = H^0 + \lambda H^{(1)}$$

where H^0 refers to the **zero order problem**, whose solution we know, and H^1 is the **perturbation**. We normally include the arbitrary parameter λ so as to keep track of orders of magnitude in the derivation.

Perturbation theory aims to write solutions for

$$H\Psi_i = \epsilon_i \Psi_i$$

in terms of the zero order problem thus

$$\psi_i = \psi_i^0 + \lambda \psi_i^{(1)} + \lambda^2 \psi_i^{(2)} + \ldots$$
$$\epsilon_i = \epsilon_i^0 + \lambda \epsilon_i^{(1)} + \lambda^2 \epsilon_i^{(2)} + \ldots$$

We call these expansions **perturbation expansions** of the wavefunction and the energy, and the result correct to second order is

$$\epsilon_i = \epsilon_i^0 + \lambda H_{ii}^{(1)} + \lambda^2 \sum_{m \neq i} \frac{H_{im}^0 H_{mi}^0}{\epsilon_i^0 - \epsilon_m^0} + \ldots$$

with a corresponding expression for ψ_i. In this expression,

$$H_{mi}^1 = \int \psi_m^0 H^{(1)} \psi_i^0 \mathrm{d}\tau$$

the idea is to take the zero order problem as a Hartree Fock model and the perturbation as the difference between the true Hamiltonian and the Hartree Fock Hamiltonian. Remember that the HF Hamiltonian averages over the electron motion.

One of the great advantages of perturbation theory is that it can be incorporated into HF codes because it does not need full transformation of the two-electron integrals. All that is needed is a partial transformation to calculate integrals such as

$$\sum \sum a_i b_j \int \int \chi_i(\mathbf{r}_1) \chi_j(\mathbf{r}_1) g(\mathbf{r}_1, \mathbf{r}_2) \psi_X(\mathbf{r}_2) \psi_Y(\mathbf{r}_2) \mathrm{d}\tau_1 \mathrm{d}\tau_2$$

which can then be stored or used as calculated.

Pople Binkley and Seeger, amongst others, have made a systematic study of the Møller Plesset method. The jargon is MP2 for a second order treatment, MP3 for a third order treatment, etc. Here is Pople's summary.

> ## Theoretical Models Incorporating Electron Correlation
> ### John A. Pople, J. Stephen Binkley and Rolf Seeger
> *International Journal of Quantum Chemistry*, Symp No 10 (1976) 1
>
> Some methods of describing electron correlation are compared from the point of view of requirements for theoretical chemical models. The perturbation approach originally introduced by Møller and Plesset, terminated at finite order, is found to satisfy most of these requirements. It is size consistent, that is applicable to an ensemble of isolated systems in an additive manner. On the other hand, it does not provide an upper bound for the electronic energy . . .
>
> Equilibrium geometries, dissociation energies, and energy separations between electronic states of different spin multiplicities are described substantially better by Møller Plesset theory to second or third order than by Hartree–Fock theory.

As a simple example, let's study the neon pair interaction potential. I have used a decent basis set, 6–311G**, and a typical output is given in the next section.

10.1.2 A MP2 GAUSSIAN92 run

Default route: SCF = Direct MP2 = Stingy MAXDISK = 25000000

```
----------------
# MP2/6-311G** density = all
----------------
----------------
Neon. . .Neon interaction energy
----------------
Symbolic Z-matrix:
   Charge = 0 Multiplicity = 1
Ne1
Ne2 Ne1 R
   Variables:
R 2.
GradGradGradGradGradGradGradGradGradGradGradGradGradGradGrad
Berny optimization.
Initialization pass.
----------------
```

Initial Parameters
(Angstroms and Degrees)

Standard
HF/6–311G** calc.

```
---------------------------------------------------------
Name Value Derivative information (Atomic Units)
---------------------------------------------------------
R 2. estimate D2E/DX2
---------------------------------------------------------
Initial trust radius is 1.000D-01.
Number of steps in this run = 20 maximum allowed number of steps = 100.
GradGradGradGradGradGradGradGradGradGradGradGradGradGradGrad
---------------------------------------------------------
```

Z-MATRIX (ANGSTROMS AND DEGREES)
CD   Cent Atom N1 Length/X N2 Alpha/Y N3 Beta/Z   J
----
1 1   Ne
2 2   Ne   1   2.000000(1)

                        Z-Matrix orientation:

Center   Atomic   Coordinates (Angstroms)
Number   Number   X   Y   Z
----
1   10   .000000   .000000   .000000
2   10   .000000   .000000   2.000000

STOICHIOMETRY   Ne2
FRAMEWORK GROUP   D*H[C*(Ne.Ne)]
DEG. OF FREEDOM   1
FULL POINT GROUP   D*H   NOP 8
LARGEST ABELIAN SUBGROUP   D2H   NOP 8
LARGEST CONCISE ABELIAN SUBGROUP   C2   NOP 2
                        Standard orientation:

Center   Atomic   Coordinates (Angstroms)
Number   Number   X   Y   Z
----
1   10   .000000   .000000   1.000000
2   10   .000000   .000000   -1.000000

Rotational constants (GHZ): .0000000 12.6392550 12.6392550
Isotopes: Ne-20,Ne-20
Standard basis: 6-311G(D,P) (S, S=P, 5D, 7F)
There are 9 symmetry adapted basis functions of AG symmetry.
There are 1 symmetry adapted basis functions of B1G symmetry.
There are 4 symmetry adapted basis functions of B2G symmetry.
There are 4 symmetry adapted basis functions of B3G symmetry.
There are 1 symmetry adapted basis functions of AU symmetry.
There are 9 symmetry adapted basis functions of B1U symmetry.
There are 4 symmetry adapted basis functions of B2U symmetry.
There are 4 symmetry adapted basis functions of B3U symmetry.
Crude estimate of integral set expansion from redundant integrals = 1.000.
Integral buffers will be 262144 words long.
Raffenetti 1 integral format.
Two-electron integral symmetry is turned on.
   36 basis functions   62 primitive gaussians
   10 alpha electrons   10 beta electrons
      nuclear repulsion energy 26.4588624500 Hartrees.
One-electron integrals computed using PRISM.
One-electron integral symmetry used in STVInt
The smallest eigenvalue of the overlap matrix is 8.585D-02
DipDrv: MaxL = 4.
DipDrv: will hold 34 matrices at once.
PROJECTED INDO GUESS.
INITIAL GUESS ORBITAL SYMMETRIES.
   OCCUPIED (?A) (?A) (?A) (?A) (?A) (?A) (?A) (?A) (?A) (?A)
   VIRTUAL (?A) (?B) (?A) (?A) (?B) (?A) (?A) (?B) (?A) (?A)
     (?A) (?C) (?C) (?A) (?B) (?A) (?A) (?B) (?A) (?A)

(?B) (?A) (?A) (?A) (?C) (?C)
Alpha deviation from unit magnitude is 1.33D-15 for orbital 11.
Alpha deviation from orthogonality is 4.25D-16 for orbitals 24 2.
A Direct SCF calculation will be performed.
Using DIIS extrapolation.
Closed shell SCF:
Requested convergence on RMS density matrix = 1.00D-08 within 64 cycles.
Requested convergence on MAX density matrix = 1.00D-06.
Two-electron integral symmetry used by symmetrizing Fock matrices.
Keep R1 integrals in memory in canonical form, NReq = 716313.
IEnd = 28575 IEndB = 28575 NGot = 2048000 MDV = 1753015
LenX = 1753015
MinBra = 0 MaxBra = 3 MinLOS = -1 MaxLOS = -1 MinRaf = 0 MaxRaf = 3 MinLRy = 4.
IRaf = 0 NMat = 1 IRICut = 1 DoRegI = T DoRafI = F ISym2E = 1 JSym2E = 1.
Fock matrices symmetrized in FoFDir.
SCF DONE: E(RHF) = -257.032317082   A.U. AFTER 8
CYCLES   **HF energy**
    CONVG = .7013D-08   -V/T = 1.9992
    S**2 = .0000
KE = 2.572307377753D + 02   PE = -6.754516870820D + 02   EE = 1.347297697747D + 02
Range of M.O.s used for correlation: 3 36
NBasis = 36 NAE = 10 NBE = 10 NFC = 2 NFV = 0
NROrb = 34 NOA = 8 NOB = 8 NVA = 26 NVB = 26
Frozen-core derivative calculation, NFC = 2 NFV = 0.
FulOut = F Deriv = T AODrv = F
MMem = 18   MDisk = 8   NDisk1 = 76665
NDisk = 1637320   MDiskD = 8   NOA = 10
W3Min = 54872   MinDsk = 107057   NDisk2 = 1167640
NBas6D = 38   NBas2D = 807   NTT = 741
NDExt = 0   LenExt = 1024000   MDV = 2048000
MDiskM = 561
Fully in-core method, ICMem = 1731581.
JobTyp = 1 Pass 1 fully in-core.
Compute canonical integrals, IntTyp = 4.
MinBra = 0 MaxBra = 3 MinLOS = -1 MaxLOS = -1 MinRaf = 0 MaxRaf = 3 MinLRy = 4.
IRaf = 0 NMat = 1 IRICut = 1 DoRegI = T DoRafI = F ISym2E = 0 JSym2E = 0.
Symmetry not used in FoFDir.
ANorm = .1027266570D + 01   **MP2 energy**
E2 = -.4178297993D + 00   EUMP2 = -.25745014688142D + 03
Spin components of T(2) and E(2):
    alpha-alpha T2 = .7522620521D-02 E2 = -.5730449014D-01
    alpha-beta T2 = .4023136539D-01 E2 = -.3032208190D + 00
    beta-beta T2 = .7522620521D-02 E2 = -.5730449014D-01
The integrals were generated 1 times.
Minotr: Closed-shell wavefunction.
    Computing MP2 derivatives.
    Using Z-Vector for PSCF gradient.
    Skipping F1 and S1 gradient terms here.
    Frozen-core window.
    Direct CPHF calculation.
    Solving linear equations simultaneously.
    Using symmetry in CPHF.
    Requested convergence is 1.0D-10 RMS, and 1.0D-09 maximum.
    Secondary convergence is 1.0D-12 RMS, and 1.0D-12 maximum.
    Differentiating once with respect to electric field.
        with respect to dipole field.

MDV = 2048000
  Store integrals in memory, NReq = 697984.
MinBra = 0 MaxBra = 3 MinLOS = -1 MaxLOS = -1 MinRaf = 0 MaxRaf = 3 MinLRy = 4.
IRaf = 0 NMat = 1 IRICut = 1 DoRegI = T DoRafI = F ISym2E = 0 JSym2E = 0.
Symmetry not used in FoFDir.
  There are 1 degrees of freedom in the 1st order CPHF.
 1 vectors were produced by pass 0.
AX will form 1 AO Fock derivatives at one time.
 1 vectors were produced by pass 1.
 1 vectors were produced by pass 2.
 1 vectors were produced by pass 3.
 1 vectors were produced by pass 4.
 1 vectors were produced by pass 5.
 1 vectors were produced by pass 6.
 1 vectors were produced by pass 7.
Inv2: IOpt = 1 Iter = 1 AM = 2.74D-16 Conv = 1.00D-12.
Inverted reduced A of dimension 8 with in-core refinement.
Copying SCF densities to generalized density rwf, ISCF = 0 IROHF = 0.

****
Population analysis using the MP2 density.

****

> **Use the MP2 density to calculate properties**

Condensed to atoms (all electrons):
           1   2
1 Ne   10.020230   -.020230
2 Ne   -.020230 10.020230
Total atomic charges:
           1
1 Ne   .000000
2 Ne   .000000
Sum of Mulliken charges = .00000
Atomic charges with hydrogens summed into heavy atoms:
           1
1 Ne   .000000
2 Ne   .000000
Sum of Mulliken charges = .00000
Electronic spatial extent (au): $<R^{**}2> = 90.0868$
Charge = .0000 electrons
Dipole moment (Debye):
X = .0000 Y = .0000 Z = .0000 Tot = .0000
Quadrupole moment (Debye-Ang):
XX = -8.3312 YY = -8.3312 ZZ = -8.4435
XY = .0000 XZ = .0000 YZ = .0000
Octapole moment (Debye-Ang**2):
XXX = .0000 YYY = .0000 ZZZ = .0000 XYY = .0000
XXY = .0000 XXZ = .0000 XZZ = .0000 YZZ = .0000
YYZ = .0000 XYZ = .0000
Hexadecapole moment (Debye-Ang**3):
XXXX = -3.6926 YYYY = -3.6926 ZZZZ = -53.8966 XXXY = .0000
XXXZ = .0000 YYYX = .0000 YYYZ = .0000 ZZZX = .0000
ZZZY = .0000 XXYY = -1.2309 XXZZ = -9.5830 YYZZ = -9.5830
XXYZ = .0000 YYXZ = .0000 ZZXY = .0000
N-N = 2.645886245000D+01 E-N = -6.752839201783D+02 KE = 2.575842369078D+02

It is of interest to study the behaviour of the interaction energy as a function of distance. In classical language, the interaction between two neon atoms would be put down to dispersion. The results are shown in Table 10.1.

Table 10.1   HF/6–311G** and MP2/6–311G** calculations on Ne...Ne

R/Å	HF energy/E_h	MP2 energy/E_h
1.0	−255.582560	−256.015929
2.0	−257.032317	−257.450147
2.2	−257.040607	−257.458719
2.4	−257.043786	−257.462009
2.6	−257.044904	−257.463112
2.8	−257.045187	−257.463334
3.0	−257.045191	−257.463286
3.2	−257.045149	−257.463214
3.4	−257.045121	−257.463173
4.0	−275.045106	−257.463148
100.	−257.045106	−257.463144
Atom only	−128.522553	−128.731572

There are several points to note, the most important being that the MP2 calculation has the 'right' behaviour at large R; the Ne...Ne energy is just twice the atomic energy. Don't confuse this with the behaviour of HF wavefunctions at large R; some methods of treating electron correlation simply do not 'scale' properly. This explains Pople and coworkers' comment that the MP method '... is size consistent'. The resultant curve is shown in Figure 10.2. The interaction is obviously a very weak one, and it would be a confident person who put a great deal of trust in the calculated value without investigating further!

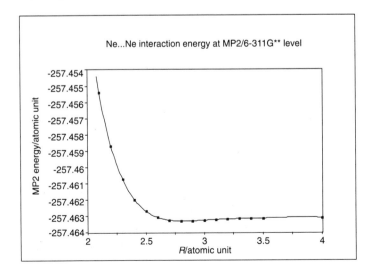

Figure 10.2   Ne...Ne interaction at MP2/5–311G** level

10.2 CONFIGURATION INTERACTION (CI)

I mentioned CI at least twice before, namely, in our discussion of H_2 and in the ZDO treatment of pyridine. CI is used in two ways:

- to improve the ground state wavefunction and/or;
- to give a description of the excited states.

In the first case, we would look for a perturbation treatment, whilst in the second case we would have to concern ourselves with the full CI Hamiltonian matrix. Perhaps the best place to start is with Foresman and coworkers' paper.

Towards a Systematic Molecular Orbital Theory for Excited States
James B. Foresman, Martin Head-Gordon, John A. Pople and Michael J. Frisch
Journal of Physical Chemistry, **96** (1992) 135

This work reviews the methodological and computational considerations necessary for the determination of the *ab initio* energy, wavefunction and gradient of a molecule in an electronically excited state using molecular orbital theory. In particular, this paper reexamines a fundamental level of theory which was employed several years ago for the interpretation of the electronic spectra of simple organic molecules: configuration interaction (CI) among all singly substituted determinants using a Hartree Fock reference state. This investigation presents several new enhancements to this general theory. First, it is shown how the CI singles wavefunction can be used to compute efficiently the analytic first derivative of the energy... Second, a computer program is described which allows these computations to be done in a 'direct' fashion,

---

Part of the paper deals with energy gradients (the first derivative of the energy with respect to bond lengths, bond angles, etc. That is a subject for chapter 12.

The basic idea is to take a single determinant HF wavefunction and to produce excited state wavefunctions from it by replacing each occupied orbital in turn by the virtual orbitals. Linear combinations of the resulting Slater determinants are then found, and these describe the ground state and the excited states. If **all possible** singly excited states are taken, then we reach a level of theory known variously as single excitation configuration interaction (SECI), monoexcited CI, or the Tamm-Dancoff approximation.

$$\Psi_0 = \psi_A^2 \psi_B^2 \dots \psi_M^2$$

The treatment is similar to that discussed within the ZDO method in Chapter 7; the ground state wavefunction is a single Slater determinant describing the lowest energy configuration. In Chapter 7, I mentioned that we would often, when considering a singly excited state, take the building blocks for such a CI calculation as the linear combination of two Slater determinants needed to make an eigenfunction of the spin operator S^2, when considering a singly excited state. I wrote this as $\Psi(A \to X)$. So, to write out the Slater determinants in full, the ground state wavefunction is

$$\Psi_0(\underline{x}_1, \underline{x}_2, \ldots \underline{x}_{2M}) = \begin{vmatrix} \psi_A(\underline{r}_1)\alpha(s_1) & \psi_A(\underline{r}_1)\beta(s_1) & \ldots \psi_M(\underline{r}_1)\beta(s_1) \\ \ldots & & \\ \psi_A(\underline{r}_{2M})\alpha(s_{2M}) & \psi_A(\underline{r}_{2M})\beta(s_{2M}) & \ldots \psi_M(\underline{r}_{2M})\beta(s_{2M}) \end{vmatrix}$$

and $\Psi(A \to X)$ is written as the difference of the two determinants

$$\begin{vmatrix} \psi_A(\underline{r}_1)\alpha(s_1) & \psi_X(\underline{r}_1)\beta(s_1) & \ldots \psi_M(\underline{r}_1)\beta(s_1) \\ \ldots & & \\ \psi_A(\underline{r}_{2M})\alpha(s_{2M}) & \psi_X(\underline{r}_{2M})\beta(s_{2M}) & \ldots psi_M(\underline{r}_{2M})\beta(s_{2M}) \end{vmatrix}$$

and

$$\begin{vmatrix} \psi_X(\underline{r}_1)\alpha(s_1) & \psi_A(\underline{r}_1)\beta(s_1) & \ldots \psi_M(\underline{r}_1)\beta(s_1) \\ \ldots & & \\ \psi_X(\underline{r}_{2M})\alpha(s_{2M}) & \psi_A(\underline{r}_{2M})\beta(s_{2M}) & \ldots \psi_M(\underline{r}_{2M})\beta(s_{2M}) \end{vmatrix}$$

In modern CI calculations, it turns out to be computationally more convenient to work with the single Slater determinants listed above, rather than take combinations at the outset. What we do is to substitute each of the $2M$ occupied spin orbitals in turn with each of the virtual spin orbitals, and then use building blocks that are **single** Slater determinants rather than simple linear combinations. Such single determinants by themselves are not necessarily, individually, eigenfunctions of the spin operator S^2, which is why it is necessary to include all of them in the CI calculation.

10.2.1 A CIS GAUSSIAN92 run

Well, here is an *ab initio* calculation on furan at the CIS/4–31G level of theory. I first optimized the geometry, using the Z-matrix discussed in Chapter 10.

```
**********************************************
Gaussian 92: HP-PARisc-HPUX-G92RevE.1 12-Jun-1993 6-Feb-1995
**********************************************
%save
Default route: SCF = Direct MP2 = Stingy MAXDISK = 25000000
----------------
# CIS(NSTATES = 10)/4-31G
----------------
FURAN
----------------
Symbolic Z-matrix:
   Charge = 0 Multiplicity = 1
C3    X    HALF34
O     X    ROX     C3    90.
C4    X    HALF34  O     90.    C3    180.    0
```

(Make sure to use C_{2v} symmetry)

C2	C3	R23	X	A23X	O	0.	0
C5	C4	R23	X	A23X	O	0.	0
H2	C2	RCH	O	AA	X	180.	0
H3	C3	RCH	C2	BB	H2	0.	0
H5	C5	RCH	O	AA	X	180.	0
H4	C4	RCH	C5	BB	O	180.	0

Variables:

HALF34	0.7219
ROX	2.0933
R23	1.3395
RCH	1.0632
A23X	106.676
AA	116.4592
BB	126.685

Z-MATRIX (ANGSTROMS AND DEGREES)

CD	Cent	Atom	N1	Length/X	N2	Alpha/Y	N3	Beta/Z	J
1	X								
2	1	C	1	.721900(1)					
3	2	O	1	2.093300(2)	2	90.000(10)			
4	3	C	1	.721900(3)	3	90.000(11)	2	180.000(18)	0
5	4	C	2	1.339500(4)	1	106.676(12)	3	.000(19)	0
6	5	C	4	1.339500(5)	1	106.676(13)	3	.000(20)	0
7	6	H	5	1.063200(6)	3	116.459(14)	1	180.000(21)	0
8	7	H	2	1.063200(7)	5	126.685(15)	7	.000(22)	0
9	8	H	6	1.063200(8)	3	116.459(16)	1	180.000(23)	0
10	9	H	4	1.063200(9)	6	126.685(17)	3	180.000(24)	0

Z-Matrix orientation:

Center Number	Atomic Number	Coordinates (Angstroms)		
		X	Y	Z
1	-1	.000000	.000000	.000000
2	6	.000000	.000000	.721900
3	8	2.093300	.000000	.000000
4	6	.000000	.000000	-.721900
5	6	1.283164	.000000	1.106282
6	6	1.283164	.000000	-1.106282
7	1	1.771218	.000000	2.050844
8	1	-.853124	.000000	1.356387
9	1	1.771218	.000000	-2.050844
10	1	-.853124	.000000	-1.356387

STOICHIOMETRY   C4H4O
FRAMEWORK GROUP   C2V[C2(O),SGV(C4H4)]
DEG. OF FREEDOM   8
FULL POINT GROUP   C2V   NOP 4
LARGEST ABELIAN SUBGROUP   C2V   NOP 4
LARGEST CONCISE ABELIAN SUBGROUP   C2   NOP 2

Standard basis: 4-31G (S, S = P, 6D, 7F)
There are   23 symmetry adapted basis functions of A1 symmetry.
There are   4 symmetry adapted basis functions of A2 symmetry.
There are   6 symmetry adapted basis functions of B1 symmetry.

There are   20 symmetry adapted basis functions of B2 symmetry.
Crude estimate of integral set expansion from redundant integrals = 1.001.
Integral buffers will be 262144 words long.
Raffenetti 1 integral format.
Two-electron integral symmetry is turned on.
  53 basis functions   116 primitive gaussians
  18 alpha electrons   18 beta electrons
  nuclear repulsion energy 161.4045411816 Hartrees.

PROJECTED INDO GUESS.
INITIAL GUESS ORBITAL SYMMETRIES.
   OCCUPIED (?A) (A1) (?A) (?A) (?A) (A1) (A1) (B2) (A1) (B2)
      (A1) (B1) (B2) (A1) (B2) (A1) (B1) (A2)
   VIRTUAL (B1) (A1) (B2) (A2) (A1) (B2) (B2) (A1) (A1) (B2)
      (B2) (A1) (?A) (A1) (A1) (A1) (?A) (A1) (A1) (A1)
      (B1) (A1) (A1) (A1) (?A) (A1) (A1) (?A) (A1) (A1)
      (A1) (A1) (A1) (?A) (?A)

Initial HF calculation

A Direct SCF calculation will be performed.
Using DIIS extrapolation.
Closed shell SCF:
SCF DONE: E(RHF) = -228.286594272   A.U. AFTER 12 CYCLES
      CONVG = .7275D-08   -V/T = 1.9992
      S**2 = .0000
KE = 2.284626206314D + 02 PE = -8.576805998965D + 02 EE = 2.395268438113D + 02

Range of M.O.s used for correlation: 6   53
Semi-Direct transformation.
(rs|ai) integrals will be sorted in core.
Spin components of T(2) and E(2):
alpha-alpha T2 =   .2327276769D-01 E2 =   -.5950076005D-01
alpha-beta T2 =   .1375669832D + 00 E2 =   -.3539842364D + 00
beta-beta T2   =   .2327276769D-01 E2 =   -.5950076005D-01

ANorm =   .1088169343D + 01
E2 =   -.4729857565D + 00 EUMP2 =   -.22875958002867D + 03
RHF GROUND STATE
MDV IS: 2048000
Making orbital integer symmetry assigments:
ORBITAL SYMMETRIES.
   OCCUPIED (A1) (B2) (A1) (A1) (B2) (A1) (A1) (B2) (B2) (A1)
      (A1) (B1) (B2) (B2) (A1) (A1) (B1) (A2)
   VIRTUAL (B1) (A2) (A1) (A1) (B2) (B2) (A1) (B2) (A1) (B2)
      (B2) (A1) (B1) (B2) (A2) (A1) (A1) (B1) (B2) (A2)
      (A1) (B2) (B2) (A1) (A1) (B2) (B1) (B2) (A1) (A1)
      (B2) (A1) (B2) (B2) (A1)
40 initial guesses have been made.
Convergence on wavefunction: .000100000000000
Davidson Disk Diagonalization is being used.
max sub-space: 200 roots to seek: 40 dimension of matrix: 910
ITERATION 1 DIMENSION IS 40
New state   3 was old state   4
New state   4 was old state   13
New state   5 was old state   6
New state   6 was old state   5
New state   9 was old state   3

New state   10 was old state   16

Excitation Energies [eV] at current iteration:

Root	1:	8.270723268212885
Root	2:	9.059595349515931
Root	3:	9.919098328552233
Root	4:	10.521327233694444
Root	5:	10.563802831634573
Root	6:	10.637517262356717
Root	7:	11.458916711684337
Root	8:	11.656582458475053
Root	9:	11.865455618275558
Root	10:	11.872955842437258
Root	11:	12.026202110437263
Root	12:	12.038130874447320
Root	13:	12.329694473261193
Root	14:	12.703465431709799
Root	15:	12.820879538478402
Root	16:	13.219721261363725
Root	17:	13.308769834893222
Root	18:	13.679141671243369
Root	19:	13.778054579394394
Root	20:	14.346682592675682
Root	21:	14.426994148774352
Root	22:	14.836810283704650
Root	23:	14.841304331191933
Root	24:	14.868884895495903
Root	25:	15.740852548456260
Root	26:	15.933116347779992
Root	27:	16.350873684603068
Root	28:	16.425580397349926
Root	29:	16.647814028292733
Root	30:	16.665039386331945
Root	31:	17.197191028907859
Root	32:	17.276196704134336
Root	33:	17.445127152257683
Root	34:	17.945164480086728
Root	35:	17.947571626306164
Root	36:	18.703249276837021
Root	37:	18.805231193616756
Root	38:	18.943784127655315
Root	39:	18.963997302916237
Root	40:	19.335562031832113

ITERATION   2 DIMENSION IS   80
Root   1 not converged, maximum delta is   .044778114580726
Root   2 not converged, maximum delta is   .074169223333817

Convergence on expansion vectors, NOT on wavefunctions!

****

EXCITED STATES FROM < AA,BB:AA,BB > SINGLES MATRIX:

****

1PDM for each excited state written to RWF 633

Ground to excited state Transition Electric Dipole Moments (Au):

state	X	Y	Z	Osc.
1	.0000	-1.0499	.0000	.1978

2	.0000	.0000	.3177	.0217
3	.0000	.0000	.0000	.0000
4	.0000	.0000	-1.8414	.8413
5	.0776	.0000	.0000	.0015
6	-.1128	.0000	.0000	.0032
7	.0000	.0000	.0000	.0000
8	.0000	.8686	.0000	.2006
9	.0155	.0000	.0000	.0001
10	.0000	.0000	.0000	.0000

Transition moments (electric velocity dipole)

Ground to excited state Transition Velocity Dipole Moments (Au):

state	X	Y	Z	Osc.
1	.0000	.0390	.0000	.0038
2	.0000	.0000	.0225	.0011
3	.0000	.0000	.0000	.0000
4	.0000	.0000	.1705	.0521
5	-.0504	.0000	.0000	.0045
6	.0776	.0000	.0000	.0105
7	.0000	.0000	.0000	.0000
8	.0000	-.1133	.0000	.0215
9	.0174	.0000	.0000	.0005
10	.0000	.0000	.0000	.0000

Transition moments (magnetic dipole)

Ground to excited state Transition Magnetic Dipole Moments (Au):

state	X	Y	Z
1	.7287	.0000	.0000
2	.0000	.0000	.0000
3	.0000	.0000	.6884
4	.0000	.0000	.0000
5	.0000	-1.8798	.0000
6	.0000	-.2309	.0000
7	.0000	.0000	.1910
8	.3244	.0000	.0000
9	.0000	-.1281	.0000
10	.0000	.0000	-.9804

Ground to excited state transition densities written to RWF 633

Excitation energies and oscillator strengths:

Excited State 1: Singlet-B2 7.3255 eV 169.25 nm f=0.1978
```
17 -> 20        .14607
18 -> 19        .67543
```
This State for optimization and/or second-order correction:
Total Energy, E(Cis) = -228.017386275
Copying the Cisingles density for this state as the 1-particle RhoCI density.

Excited State 2: Singlet-A1 8.7599 eV 141.53 nm f=0.0217
```
17 -> 19        .57937
18 -> 20        -.38713
```

Excited State 3: Singlet-A2 9.7115 eV 127.67 nm f=0.0000
```
17 -> 23        .10915
17 -> 24        .10904
18 -> 21        -.61538
18 -> 22        .17520
18 -> 25        .22989
```

Excited State 4: Singlet-A1 10.1269 eV 122.43 nm f = 0.8413
    17 –> 19   -.36490
    18 –> 20   -.56616

Excited State 5: Singlet-B1 10.1388 eV 122.29 nm f = 0.0015
    14 –> 20   -.15662
    15 –> 19   -.13518
    16 –> 19   .66122

Excited State 6: Singlet-B1 10.3833 eV 119.41 nm f = 0.0032
    17 –> 21   -.33246
    18 –> 23   .53045
    18 –> 24   .13557
    18 –> 26   -.24376

Excited State 7: Singlet-A2 10.4338 eV 118.83 nm f = 0.0000
    18 –> 21   .25928
    18 –> 22   .56729
    18 –> 25   .28714

Excited State 8: Singlet-B2 10.8499 eV 114.27 nm f = 0.2006
    17 –> 20   .67218
    18 –> 19   -.14303

Excited State 9: Singlet-B1 11.2031 eV 110.67 nm f = 0.0001
    15 –> 19   .11049
    17 –> 21   -.32017
    17 –> 22   .30949
    18 –> 23   -.30518
    18 –> 24   .41694

Excited State 10: Singlet-A2 11.4671 eV 108.12 nm f = 0.0000
    9 –> 19   .13043
    11 –> 20   -.10631
    13 –> 19   -.16228
    14 –> 19   -.46771
    15 –> 20   .11950
    16 –> 20   .41029
    18 –> 25   -.13972

Job cpu time: 0 days 0 hours 1 minutes 22.7 seconds.
File lengths (MBytes): RWF = 43 Int = 0 D2E = 0 Chk = 2 Scr = 0
Normal termination of Gaussian 92.

10.3 CID* AND CISD†

The easiest way to treat electron correlation in a ground-state calculation is at the MPn‡ level of theory. The model is cheap in computer resource, primarily because it is not necessary to transform the two-electron integrals completely from the basis set to the LCAO-MO set. It **can** be a resource consuming technique in terms of disk space, but direct methods have been developed to get over this particular problem.

*MPn   Møller-Plesset (n[th] level of perturbation theory)
†CID   Configuration Interaction (double excitations)
‡CISD   Configuration Interaction (single and double excitations)
        MP = generic name for MPn.

Most modern *Ab Initio* packages are equally at home when treating the ground electronic states of large molecules and when dealing with excited states. In a CID calculation, we choose a range of orbitals (occupied and virtual). For instance, it is usual to ignore all inner shells. All doubly excited configurations within the range are then taken into account in the variational calculation. In a CISD calculation, all single excitations are also added to the expansion.

10.4 RESOURCE CONSUMPTION

Many authors have drawn your attention to the cost of some of the major steps in LCAO-MO calculations. In this text, I have mentioned HF, MP calculations and simple CI methodologies. The best review to my mind is that of Schlegel and Frisch (1990).

I discussed the resource requirements of HF calculations in Chapter 5. To give a flavour of the considerations needed for correlated wavefunctions, consider the integral transformation problem discussed above. The difficulty is with the transformation of the two-electron integrals from an atomic orbital basis to a MO basis. This requires roughly the work needed to transform n(LCAO-MOs)$\times n^4$ atomic orbitals into two-electron integrals over the LCAO coefficients. There are several ways of carrying out this transformation depending on how the atomic integrals are stored. If the integrals are stored in the so-called **canonical order** (McWeeny and Sutcliffe 1969), then the calculation can be vectorized for a vector processor. Normally, zero integrals are not stored and integrals are written in the order that they are calculated. This often implies the use of molecular symmetry.

A small subset of Schlegel and Frisch's table is shown below (Table 10.2).

Table 10.2 Cost dependence for some popular *ab initio* models. n = no. of basis functions; O = no. of occupied MOs

	CPU usage	Disk usage
Conventional HF	$n^{3.5}$	$n^{3.5}$
Direct HF	$n^{2.7}$	n^2
Conventional MP2	On^4	n^4
Direct MP2	$O^2 n^3$	n^2
CISD	$O^2 n^4$	n^4

11   The Xα Model

I have dealt at length with the Hartree Fock model but I haven't said much in this book so far about atoms. Hartree was actually concerned with the **atomic** problem, which he envisaged as a kind of one-electron problem. He focused attention on one (**any** one) of the electrons, and sought to replace the complicated field produced by the other electrons, which depends on just where those electrons are, by a simple model potential arising from their average positions. He achieved this by assigning to each electron in an atom its own individual orbital and energy. Hartree also assumed that the many-electron wavefunction for the atom could be written as a simple product of orbitals. Fock pointed out shortly afterwards that one could take determinantal wavefunctions in order to accommodate the exclusion principle.

I did mention earlier that the theory of atomic structure is very much a subject in its own right and you will find it well worth your while reading the two classic books by Hartree (1957) and Fischer (1977) on the atomic problem.

The atomic problem is quite different from the molecular one, because of spherical symmetry. The theory is simplified (or dominated, depending on your viewpoint) by angular momentum and the Hartree Fock limit can be reached easily without recourse to the LCAO approximation. Obviously, from what I have already said, a great deal of effort has also gone into LCAO expansions for atoms. There has been a resurgence of interest recently in atomic SCF calculations because astrophysicists want to study highly ionized atomic species in the interstellar medium, and they look to theory for their energy level data rather than earth-bound experiments where the species are hard to produce.

11.1   THE ATOMIC HF PROBLEM

In Chapter 5, I focused attention on molecular systems with singlet electronic ground states, and in particular those that can be written with molecular orbital configurations typified by $\psi_A^2 \psi_B^2 \ldots \psi_M^2$, and I mentioned the 'analytical' treatment of the self consistent field (i.e. the one before invoking the LCAO approximation). To remind you, what we do is this.

- In the language of Slater determinants, the smallest possible logical building block for such a system is

$$\Psi_{el}(\underline{x}_1, \underline{x}_2, \ldots \underline{x}_{2M}) = \begin{vmatrix} \psi_A(\mathbf{r}_1)\alpha(s_1) & \psi_A(\mathbf{r}_1)\beta(s_1) \ldots \psi_M(\mathbf{r}_1)\beta(s_1) \\ \cdots \\ \psi_A(\mathbf{r}_{2M})\alpha(s_{2M}) & \psi_A(\mathbf{r}_{2M})\beta(s_{2M}) \ldots \psi_M(\mathbf{r}_{2M})\beta(s_{2M}) \end{vmatrix}$$

where we have taken account of antisymmetry by writing a Slater determinant.
- The aim of SCF theory is to find the best form of the one-electron functions by minimizing the variational energy

$$\epsilon_{el} = \frac{\int \Psi_{el} H_{el} \Psi_{el} d\tau}{\int \Psi_{el}^2 d\tau}$$

This integration is over the coordinates of all the electrons, and I have assumed that all the values of ψ are real quantities rather than complex ones. I showed you how to do it, in principle; we
- work out the formal energy expression, for which we find

$$\epsilon_{el} = 2 \sum_{R=A}^{M} \int \psi_R(\mathbf{r}_1) h(\mathbf{r}_1) \psi_R(\mathbf{r}_1) d\tau_1$$

$$+ \sum_{R=A}^{M} \sum_{S=A}^{M} [2 \int \int \psi_R(\mathbf{r}_1) psi_R(\mathbf{r}_1) g(\mathbf{r}_1, \mathbf{r}_2) \psi_S(\mathbf{r}_2) \psi_S(\mathbf{r}_2) d\tau_1 d\tau_2$$

$$- \int \int psi_R(\mathbf{r}_1) \psi_S(\mathbf{r}_1) g(\mathbf{r}_1, \mathbf{r}_1) \psi_R(\mathbf{r}_2) \psi_S(\mathbf{r}_2) d\tau_1 d\tau_2]$$

- and then vary the ψ's in order to minimize ϵ_{el}.

I also told you that it was normal at this stage to invoke the LCAO expansion method in a molecular calculation. Forget about LCAO and let's take Hartree's formal argument to its conclusion. When we go through all the algebra, which I am not proposing to do, we arrive at the eigenvalue equation for the SCF orbitals

$$h^F(\mathbf{r})\psi_R(\mathbf{r}) = \epsilon_R \psi_R(\mathbf{r})$$

where h^F is the Hartree fock operator, which depends on the coordinates $\mathbf{r}$ of any of the electrons, and we are interested in the M lowest energy solutions. The index R therefore runs from 1 to M (the number of electron pairs). Solutions with an index greater than M certainly exist (the 'virtual' orbitals), but they don't have any useful physical interpretation.

The eigenvalue equation is often written more explicitly

$$\left\{ h(\mathbf{r}) + \sum_{S=1}^{M} [2J_S(\mathbf{r}) - K_S(\mathbf{r})] \right\} \psi_R(\mathbf{r}) = \epsilon_R \psi_R(\mathbf{r})$$

where $h(\underline{r})$ is the usual one-electron operator (the sum of the kinetic energy and the nuclear attractions for any one of the electrons), and the J and K operators are referred to as 'coulomb' and 'exchange' operators which are defined in a seemingly complicated and formal manner as follows;

$$\int \psi_R(\underline{r}_1) J_S(\underline{r}_1)\psi_R(\underline{r}_1)d\tau_1 = \int\int \psi_R(\underline{r}_1)\psi_R(\underline{r}_1)g(\underline{r}_1,\underline{r}_2)\psi_S(\underline{r}_2)\psi_S(\underline{r}_2)d\tau_1 d\tau_2$$

and

$$\int \psi_R(\underline{r}_1) K_S(\underline{r}_1)\psi_R(\underline{r}_1)d\tau = \int\int \psi_R(\underline{r}_1)\psi_S(\underline{r}_1)g(\underline{r}_1,\underline{r}_2)\psi_R(\underline{r}_2)\psi_S(\underline{r}_2)d\tau_1 d\tau_2$$

The coulomb operator is easy to understand, whilst the 'exchange' terms only arise on account of the antisymmetry principle. In Hartree's original work, he did not take antisymmetry into account and so he investigated a wavefunction (the **Hartree model**) that was just a **product** of the individual orbitals

$$\psi_{el}^{Hartree}(\underline{x}_1, \underline{x}_2, \ldots \underline{x}_{2M}) = \psi_A(\underline{r}_1)\alpha(s_1)\psi_A(\underline{r}_2)\beta(s_2)\ldots\psi_M(\underline{r}_{2M})\beta(s_{2M})$$

which gives a different energy

$$\epsilon_{el}^{Hartree} = 2\sum_{R=A}^{M}\int \psi_R(\underline{r}_1)h(\underline{r}_1)\psi_R(\underline{r}_1)d\tau_1$$

$$+ \sum_{R=A}^{M}\sum_{S=A}^{M} 2\int\int \psi_R(\underline{r}_1)\psi_R(\underline{r}_1)g(\underline{r}_1,\underline{r}_2)\psi_S(\underline{r}_2)\psi_S(\underline{r}_2)d\tau_1 d\tau_2$$

The fact of the matter is that the Hartree Fock model, which takes account of the indistinguishability of electrons, gives a very different energy from the simple Hartree model, and it turns out that this HF energy was in much better agreement with experimental results. There was initially a great deal of confusion about the extra terms in the Hartree Fock model, compared to the Hartree model. Some authors tried to ascribe a physical process to the 'exchange' terms, imagining that electrons were being 'exchanged' perhaps by some mysterious force, and attention focused on attempts to explain the 'exchange' phenomenon. More importantly, attention was devoted to finding an effective model potential which mimicked exchange.

In the early days of theoretical chemistry, researchers were interested particularly in modelling the solid state (and of course they still are). In order to give you the background to the Xα method, Let me describe a couple of the very simple models that were used many years ago, in order to understand the behaviour of electrons in metallic conductors.

11.2 MODELLING METALLIC CONDUCTORS

The simplest picture of a metallic conductor is one where we have a rigid lattice of metal (M) cations, each of which has lost one or more electrons to the surrounding 'sea' of electrons. So for a 'slice' through such a metallic conductor we might expect to see the effect illustrated in Figure 11.1, where the black 'blobs' represent the cations M^{n+}. I have deliberately drawn them as large objects, not because they are actually large, but for two different reasons. Firstly, the early models of metallic electrons such as the Drude model tried to treat these metallic electrons as ideal gases, until it was realized that the electrons collided with the cations and with each other an uncomfortably large number of times. In any case, many of the deductions from the Drude model were demonstrably flawed. Secondly, the influence of the cations on their surrounding space is not negligible. The cations produce strong electrostatic fields, and the electron density in this model is far from constant or even slowly varying with distance from the cations. Near the cationic nuclei, the electrons experience a very strong attraction.

Cations + sea of mobile electrons

Figure 11.1   Simple model of a metal

Be that as it may, the next level of model (Pauli's electron gas model) attempted to treat the electrons as if they were a gas confined to the volume of the metallic sample, but the analysis was performed quantum mechanically. The electrons are treated as if they are constrained to move in a macroscopic 'box', which for the sake of argument we can take to be a cube of side A. Inside this box, the potential is taken to be a constant U_0, and to simplify the argument we will take this constant as the zero of energy. We focus attention on a single electron moving in the box; and the electronic Schrödinger equation is

$$\left(\frac{-h^2}{8\pi^2 m}\nabla^2 + V\right)\psi = \epsilon\psi$$

where, of course, the potential V is infinite for regions outside the box and zero for regions inside the box. It is shown in many of the elementary physical chemistry textbooks that solutions of this equation are

$$\psi_{n,k,l} = \left(\sqrt{2/A}\right)^3 \sin\left(\frac{n\pi}{A}x\right) \sin\left(\frac{k\pi}{A}y\right) \sin\left(\frac{l\pi}{A}z\right)$$

$$\epsilon_{n,k,l} = \left(n^2 + k^2 + l^2\right)\frac{h^2}{8mA^2}$$

where $n, k, l = 1, 2, 3, \ldots$

There are many degeneracies, so I had better be careful to distinguish between energy levels and quantum states. Each energy level can correspond to a number of different quantum states. As the quantum numbers increase, the levels crowd closer and closer together. Not only that, but a pencil and paper calculation will reveal that very large quantum numbers (typically 10^9) are to be expected. Because the quantum states crowd so closely together, they essentially form a continuum rather than a set of discrete levels, and we focus attention on the so-called **density of states**, $D(\epsilon)$; $D(\epsilon)$ is the number of quantum states having energy between ϵ and $\epsilon + d\epsilon$, and it turns out that

$$D(\epsilon) = \frac{4\pi A^3}{h^3}(2m)^{\frac{3}{2}}\epsilon^{\frac{1}{2}}$$

11.2.1 Pauli's model

In Pauli's model, the conduction electrons are taken to be non-interacting, and so the total wavefunction for these electrons is a product of individual one-electron wavefunctions. But the Pauli model **does** take account of the exclusion principle. Each conduction electron has spin and so each available spatial quantum state can accomodate a pair of electrons, one of either spin.

    Imagine that the temperature is now fixed at $0\,\mathrm{K}$, and so the electrons have to be arranged in the available quantum states in order to produce the lowest possible energy. We therefore allocate pairs of electrons in turn into the spatial quantum states, starting with the lowest energy one, and the highest occupied level is referred to as the **Fermi level**. The energy of this state is given the symbol ϵ_F, and is called the **Fermi energy**. The number N of conduction electrons can be related to ϵ_F and we find

$$\epsilon_F = \frac{h^2}{8m}\left(\frac{3N}{A^3\pi}\right)^{\frac{2}{3}}$$

Now, N/A^3 is the number density of conduction electrons, so Pauli's model gives a simple relationship between energy and number density of electrons ρ_0. Unfortunately, people use the same symbol for a density and for a charge distribution, but I am afraid that life is like that. The electron density is $(-e)$ times the number density.

$$\epsilon_F = \frac{h^2}{8m}\left(\frac{3\rho_0}{\pi}\right)^{\frac{2}{3}}$$

ρ_0 is a constant in this simple treatment, which has dealt so far with non-interacting electrons.

11.2.2 The Thomas–Fermi model

We now switch on a potential $V(\underline{r})$ that is slowly varying over the dimensions of the box, and so the conduction electron density will become inhomogeneous. A little analysis suggests that

$$\rho(\underline{r}) = \frac{8\pi}{3h^3}(2m)^{\frac{3}{2}}[\epsilon_F - V(\underline{r})]^{\frac{3}{2}}$$

which is called the Thomas–Fermi relation. It relates the local electron density (at point $\underline{r}$) to the potential at that point.

Recall that our treatment so far has dealt with non-interacting electrons, whilst we know for sure that electrons do interact with each other. The Thomas–Fermi equation has been used in attempts to find **effective** one-body potentials in crystals, which model the effect of the other electrons. It is an interesting approach to follow because, in principle, we could measure the electron density $\rho(\underline{r})$ at points in the metallic conductor and then find the form of the potential.

Dirac (1930) studied the effects of exchange interactions in the Thomas–Fermi theory, and he discovered that the effect of exchange could be modelled by adding an extra term

$$V_X = C\rho^{\frac{1}{3}}$$

where C is a constant given by $-(e^2/4\pi\epsilon_0)(3/\pi)^{1/3}$.

$$V_X = C\rho^{\frac{1}{3}}$$

This result was later rediscovered by Slater (1951) with a slightly different numerical coefficient of $2/3\,C$. Authors often refer to a term proportional to $\rho^{1/3}$ in the one-body potential as the Dirac–Slater exchange potential V_X.

Let me just remind you that the density ρ and therefore the exchange potential both vary depending on their position in space. We often refer to models like this as 'local density' models.

The fact that Slater and Dirac got different numerical coefficients for the $\rho^{1/3}$ term was quickly resolved, and authors began to write the exchange potential as

$$V_{X\alpha} = \alpha C\rho^{\frac{1}{3}}$$

where, in principle, α should take values between $\frac{2}{3}$ and 1. I will refer to it as the $X\alpha$ exchange potential, and the idea is that you add it to a Hartree model in order to take account of antisymmetry.

11.3 SLATER'S Xα METHOD FOR ATOMS

By analogy with solid-state studies, Slater had the idea of writing the atomic Hartree Fock Hamiltonian equation

$$h^F(\mathbf{r})\psi_R(\mathbf{r}) = \epsilon_R\psi_R(\mathbf{r})$$

as

$$(h + V_C(\mathbf{r}) + V_{X\alpha}(\mathbf{r}))\psi_R(\mathbf{r}) = \epsilon_R\psi_R(\mathbf{r})$$

where h is the usual one-electron operator and V_C the coulomb operator discussed earlier. This suggestion does not make the atomic problem easier to solve and it begs the question as to which value of α should be chosen. The values given by Schwarz (1972) are normally used. Atomic Xα calculations are straightforward and most workers use the computer code due to Herman and Skillman 1963.

You will probably be interested to learn that Koopmans' theorem is not valid in its simple form for Xα wavefunctions, and so Xα SCF orbital energies do not give an estimate of the ionization energy directly. We think in terms of the **occupation number** n. In standard Hartree Fock theory, we deal with doubly occupied, singly occupied and virtual (empty) orbitals, and these orbitals have occupation numbers of 2, 1 and 0 respectively. In solid-state theory, it is conventional to think about the occupation number as if it were a continuous variable having values between 0 and 2, and I will write it as n.

In fact, in Xα theories,

$$\frac{\partial\epsilon_{el}}{\partial n_i} = \epsilon_i$$

so that the ionization energy from a particular orbital i, assuming that all the others remain constant, is given by

$$\epsilon_{el}(n_i = 0) - \epsilon_{el}(n_i = 1) = \int_1^0 \epsilon_i dn_i$$

If we make the assumption that the total .energy is a quadratic function of the occuption number, then a quick calculation shows that the ionization energy is given by the orbital energy when that orbital is half occupied.

A separate SCF calculation is therefore needed for each ionization energy, but electron reorganization can now take place. What we do is place half an electron in the orbital from which the electron is supposedly ionized, and re-do the SCF calculation. The hypothetical state with a fractional electron is called a **transition state** and we can treat the transition state according to UHF or ROHF type theories.

11.4 SLATER'S MULTIPLE SCATTERING Xα METHOD FOR MOLECULES

For an isolated atom, the Xα SCF model can be easily and quickly solved because of the spherical symmetry. By themselves, Xα calculations alone are not suitable for molecules for just the same reason; molecules do not have spherical symmetry. The molecular version of the atomic Xα method builds on the chemist's intuitive idea that a molecule is really a collection of slightly modified atoms, so for H_2O (Figure 11.2) we picture the three atoms as spheres, and inside these spheres we solve the Xα atomic problem. I have denoted the atoms as 'region I' in the figure. It turns out to be necessary to enclose the molecule in an outer sphere, and once again we solve the atomic Xα problem in region III.

In the intersphere regions, we assume a constant potential. The wavefunction and its first derivative have both got to be continuous at the points in space where the regions meet, and imposing this physical equirement gives an eigenvalue problem. The molecular Xα method has been well reviewed elsewhere, but you might like to see how it works for H_2O.

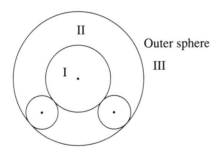

Figure 11.2   A Muffin tin model of H_2O ready for an Xα calculation

11.4.1   A simple Xα calculation

The first step is to choose a geometry, the sphere radii and the exchange parameters α for the sphere regions (Table 11.1).

Table 11.1 H_2O molecular Xα calculation. $R_{OH} = 95.84\,pm$, HOH $= 104.5°$

Atom	Sphere radius/a_0	α
H	0.310126	0.97804
O	1.498674	0.74367
Outer sphere	1.903608	0.89992

The ratio of the sphere radii were taken from Slater's (1965) table based on crystal data. The basic molecular Xα equations were originally derived on the assumption that the spheres did not overlap (Schwarz and Connolly, 1971). But the equations remain valid when the spheres are allowed to overlap, provided that each atomic sphere does not contain more than one nucleus and that none of the nuclei are outside the outer sphere. A 10% overlap seems to be normal practice. Table 11.2 gives our results.

Table 11.2　H_2O calculation, ROHF transition state. Ionization energies/E_h)

Orbital	Touching spheres	Overlap spheres	ESCA	Conventional SCF
$1b_1$	0.586	0.585	0.464	0.396
$3a_1$	0.606	0.599	0.540	0.458
$1b_2$	0.622	0.667	0.682	0.621
$2a_1$	1.149	1.140	1.181	1.277
O(1s) $1a_1$	20.10	20.10	19.890	20.429

11.5　A MODERN IMPLEMENTATION

The latest version of GAUSSIAN, GAUSSIAN94, allows for so-called density functional calculations on polyatomic molecules. The days of the muffin tin approximation have long gone, and so it is not necessary to consider sphere radii, atomic exchange factors and the like. In Chapter 5, I showed you how the electronic energy could be calculated for a wavefunction built from doubly occupied orbitals. Perhaps the most useful expression for the present is

$$\epsilon_{el} = \text{trace}\,(\underline{\mathbf{h}}_1\underline{\mathbf{P}}_1) + \frac{1}{2}\text{trace}\,(\underline{\mathbf{P}}_1\underline{\mathbf{G}})$$

The electron repulsion (G) is defined in terms of the electron density $\mathbf{P}_1$ and the two-electron integrals, and it has a complicated form

$$G_{ij} = \sum\sum P_{1,kl}\left(\int\int \chi_i(\mathbf{r})\chi_j(\mathbf{r})g(\mathbf{r}_1,\mathbf{r}_2)\chi_k(\mathbf{r}_2)\chi_l(\mathbf{r}_2)d\tau_1 d\tau_2\right.$$
$$-\frac{1}{2}\int\int \chi_i(\mathbf{r}_1)\chi_k g(\mathbf{r}_1,\mathbf{r}_2)\chi_j(\mathbf{r})\chi_l(\mathbf{r})d\tau_1 d\tau_2$$

The repulsion martix G is often written as

$$\underline{\mathbf{G}} = \underline{J} - \frac{1}{2}\underline{K}$$

where J is referred to as the **Coulomb** matrix and K the **exchange** matrix. The energy equation thus becomes

$$\epsilon_{el} = \text{trace}\,(\underline{\mathbf{h}}_1\underline{\mathbf{P}}_1 + \frac{1}{2}\text{trace}\,(\underline{\mathbf{P}}_1\underline{J}) - \frac{1}{4}\text{trace}\,(\underline{\mathbf{P}}_1\underline{K})$$

I will leave you to work out the elements of J and K for yourself, it isn't important for our discussion.

In density functional theory, we can start from the energy expression given above, and then generalize it as follows.

$$\epsilon_{el} = \text{trace}\,(\underline{\mathbf{h}}_1\underline{\mathbf{P}}_1) + \frac{1}{2}\,\text{trace}\,(\underline{\mathbf{P}}_1\underline{\mathbf{J}}) + \epsilon_X + \epsilon_C$$

Here, ϵ_X is the exchange term (which depends on the electron density, and so varies from place to place), and ϵ_C is the correlation term (which is zero in Hartree Fock theory).

In the GAUSSIAN94 implementation, you can choose from the Slater, $X\alpha$ and Becke88 models for ϵ_X, and for a variety of ϵ_C's.

To give a concrete example, here is a DFT (density functional theory) calculation on H_2O using the $X\alpha$ (exchange) and the LYP (Lee, Yang and Parr) terms.

The Input should be familiar, by now. The Schrödinger equation is integrated directly.

```
#XALYP/4-31G POP = Full

WATER MOLECULE Density Functional calculation

0   1
O
H1   0   ROH
H2   0   ROH   H1   ANG


ROH   0.956
ANG   104.5
```

And the output should be also familiar to GAUSSIAN watchers. I have given the literature citation in full, as it is different from the GAUSSIAN94 version.

A density functional calculation adds an additional step to each major phase of a Hartree Fock calculation. In particular, the accuracy depends on the number of points chosen in the numerical integration. This means that the final energy has to be used with some care.

Cite this work as:
Gaussian 94, Revision B.1,
M. J. Frisch, G. W. Trucks, H. B. Schlegel, P. M. W. Gill, B. G. Johnson, M. A. Robb, J. R. Cheeseman, T. Keith, G. A. Petersson, J. A. Montgomery, K. Raghavachari,
M. A. Al-Laham, V. G. Zakrzewski, J. V. Ortiz, J. B. Foresman, J. Cioslowski, B. B. Stefanov, A. Nanayakkara, M. Challacombe, C. Y. Peng, P. Y. Ayala, W. Chen, M. W. Wong, J. L. Andres, E. S. Replogle, R. Gomperts, R. L. Martin, D. J. Fox,
J. S. Binkley, D. J. Defrees, J. Baker, J. P. Stewart, M. Head-Gordon, C. Gonzalez, and J. A. Pople,
Gaussian, Inc, Pittsburgh PA, 1995.
* *
Gaussian 94: HP-PARisc-HPUX-G94 4evB.1 16-Apr-1995
            1-Jun-1995

* *

%save
Default route: SCF = Direct MP2 = Stingy MAXDISK = 25000000

#XALYP/4-31G POP = Full

1/38 = 1/1;
2/12 = 2,17 = 6,18 = 5/2;
3/5 = 1,11 = 2,25 = 1,30 = 1/1,2,3;
4/1;
5/5 = 2,32 = 1,38 = 4,42 = 32/2;
6/7 = 3,19 = 1,28 = 1/1;
99/5 = 1,9 = 1/99;

WATER MOLECULE

Symbolic Z-matrix:
   Charge = 0 Multiplicity = 1
O
H1   O   ROH
H2   O   ROH   H1   ANG
   Variables:
ROH   0.956
ANG   104.5

            Z-MATRIX (ANGSTROMS AND DEGREES)
CD Cent Atom N1 Length/X N2 Alpha/Y N3 Beta/Z J
----
1   1 O
2   2 H   1   0.956000(1)
3   3 H   1   0.956000(2) 2 104.500(3)

                 Z-Matrix orientation:

Centre   Atomic   Coordinates (Angstroms)
Number   Number   X   Y   Z
----
   1   8   0.000000   0.000000   0.000000
   2   1   0.000000   0.000000   0.956000
   3   1   0.925549   0.000000 -0.239363

              Distance matrix (angstroms):
                1   2   3
1 O   0.000000
2 H   0.956000   0.000000
3 H   0.956000   1.511798   0.000000

Interatomic angles:

H2-O1-H3 = 104.5
Stoichiometry   H2O
Framework group   C2V[C2(O),SGV(H2)]
Deg. of freedom   2
Full point group   C2V   NOp   4
Largest Abelian subgroup   C2V   NOp   4
Largest concise Abelian subgroup   C2   NOp   2
                         Standard orientation:

Center Number	Atom Number	Coordinates (Angstroms)		
		X	Y	Z
1	8	0.000000	0.000000	0.117056
2	1	0.000000	0.755899	-0.468224
3	1	0.000000	-0.755899	-0.468224

Rotational constants (GHZ): 824.1771445 438.8077671 286.3496857
Isotopes: O-16,H-1,H-1
Standard basis: 4-31G (6D, 7F)
There are   7 symmetry adapted basis functions of A1 symmetry.
There are   0 symmetry adapted basis functions of A2 symmetry.
There are   2 symmetry adapted basis functions of B1 symmetry.
There are   4 symmetry adapted basis functions of B2 symmetry.
Crude estimate of integral set expansion from redundant integrals = 1.247.
Integral buffers will be   262144 words long.
Raffenetti 2 integral format.
Two-electron integral symmetry is turned on.
   13 basis functions   28 primitive gaussians
   5 alpha electrons   5 beta electrons
      nuclear repulsion energy   9.2065546047 Hartrees.
One-electron integrals computed using PRISM.
The smallest eigenvalue of the overlap matrix is 7.236D-02
Projected INDO Guess.
Initial guess orbital symmetries:
   Occupied (A1) (A1) (B2) (A1) (B1)
   Virtual (A1) (B2) (A1) (B1) (B2) (A1) (A1) (B2)
Warning! Cutoffs for single-point calculations used.
Requested convergence on RMS density matrix = 1.00D-04 within 64 cycles.
Requested convergence on MAX density matrix = 1.00D-02.
Requested convergence on energy = 5.00D-05.
Keep R1 and R2 integrals in memory in canonical form, NReq = 426998.
SCF Done: E(RXa-LYP) = -75.8212722437 A.U after 5 cycles
      Convg = 0.7267D-04
-V/T = 1.9980
      $S^{**}2$ = 0.0000

****
            Population analysis using the SCF density.
****

Orbital Symmetries:
   Occupied (A1) (A1) (B2) (A1) (B1)
   Virtual (A1) (B2) (B2) (A1) (B1) (A1) (B2) (A1)
The electronic state is 1-A1.
Alpha occ. eigenvalues — 18.60697 -0.91766 -0.47215 -0.28647 -0.22395
Alpha virt. eigenvalues – 0.03593   0.12498   0.78796   0.84876   0.89197
Alpha virt. eigenvalues – 0.96460   1.06423   1.47825
Molecular Orbital Coefficients

		1	2	3	4	5
		(A1)–O	(A1)–O	(B2)–O	(A1)–O	(B1)–O
EIGENVALUES –		18.60697	-0.91766	-0.47215	-0.28647	-0.22395
1 1	O 1S	0.99036	-0.20688	0.00000	-0.09633	0.00000
2	2S	0.06104	0.41490	0.00000	0.18900	0.00000
3	2PX	0.00000	0.00000	0.00000	0.00000	0.62811
4	2PY	0.00000	0.00000	0.51131	0.00000	0.00000

5		2PZ	-0.00289	-0.16690	0.00000	0.52997	0.00000
6		3S	-0.03171	0.47832	0.00000	0.40436	0.00000
7		3PX	0.00000	0.00000	0.00000	0.00000	0.52391
8		3PY	0.00000	0.00000	0.24883	0.00000	0.00000
9		3PZ	0.00545	-0.08504	0.00000	0.39152	0.00000
10	2 H	1S	0.00052	0.14999	0.25876	-0.13127	0.00000
11		2S	0.00687	0.00435	0.15589	-0.10527	0.00000
12	3 H	1S	0.00052	0.14999	-0.25876	-0.13127	0.00000
13		2S	0.00687	0.00435	-0.15589	-0.10527	0.00000

			6	7	8	9	10
			(A1)–V	(B2)–V	(B2)–V	(A1)–V	(B1)–V
EIGENVALUES –			0.03593	0.12498	0.78796	0.84876	0.89197
1	1 O	1S	-0.09132	0.00000	0.00000	0.03555	0.00000
2		2S	0.15067	0.00000	0.00000	-0.19284	0.00000
3		2PX	0.00000	0.00000	0.00000	0.00000	-0.97160
4		2PY	0.00000	-0.42936	-0.23067	0.00000	0.00000
5		2PZ	-0.29667	0.00000	0.00000	0.72760	0.00000
6		3S	1.04789	0.00000	0.00000	0.23528	0.00000
7		3PX	0.00000	0.00000	0.00000	0.00000	1.03153
8		3PY	0.00000	-0.71074	-0.44629	0.00000	0.00000
9		3PZ	-0.41373	0.00000	0.00000	-0.38682	0.00000
10	2 H	1S	-0.10045	0.10728	0.976247	0.00000	
11		2S	-0.89946	1.23815	-0.73048	-0.59328	0.00000
12	3 H	1S	-0.10045	-0.10728	-0.97896	0.76247	0.00000
13		2S	-0.89946	-1.23815	0.73048	-0.59328	0.00000

			11	12	13
			(A1)–V	(B2)–V	(A1)–V
EIGENVALUES –			0.96460	1.06423	1.47825
1	1 O	1S	0.07530	0.00000	0.04708
2		2S	-0.40616	0.00000	-1.69102
3		2PX	0.00000	0.00000	0.00000
4		2PY	0.00000	-0.97432	0.00000
5		2PZ	-0.65006	0.00000	0.16314
6		3S	-0.02870	0.00000	2.69201
7		3PX	0.00000	0.00000	0.00000
8		3PY	0.00000	1.66773	0.00000
9		3PZ	1.17898	0.00000	-0.80327
10	2 H	1S	0.65679	-0.08776	-0.46098
11		2S	-0.12686	-0.85899	-0.57673
12	3 H	1S	0.65679	0.08776	-0.46098
13		2S	-0L.12686	0.85899	-0.57673

DENSITY MATRIX.

			1	2	3	4	5
1	1 O	1S	2.06577				
2		2S	-0.08717	0.42318			
3		2PX	0.00000	0.78905			
4		2PY	0.00000	0.00000	0.00000	0.52287	
5		2PZ	-0.03876	0.06148	0.00000	0.00000	0.61747
6		3S	-0.33861	0.54589	0.00000	0.00000	0.26912
7		3PX	0.00000	0.00000	0.65815	0.00000	0.00000
8		3PY	0.00000	0.00000	0.00000	0.25446	0.00000
9		3PZ	-0.02944	0.07809	0.00000	0.00000	0.44335
10	2 H	1S	-0.03574	0.07491	0.00000	0.26461	-0.18921
11		2S	0.03208	-0.03534	0.00000	0.15942	-0.11308
12	3 H	1S	-0.03574	0.07491	0.00000	-0.26461	-0.18921
13		2S	0.03208	-0.03534	0.00000	-0.15942	-0.11308

		6	7	8	9	10
6	3S	0.78660				
7	3PX	0.00000	0.54897			
8	3PY	0.00000	0.00000	0.12383		
9	3PZ	0.23493	0.00000	0.00000	0.32110	
10 2 H 1S		0.03729	0.00000	0.12877	-0.12830	0.21338
11	2S	-0.08141	0.00000	0.07758	-0.08310	0.10963
12 3 H 1S		0.03729	0.00000	-0.12877	-0.12830	-0.05446
13	2S	-0.08141	0.00000	-0.07758	-0.08310	-0.05173

		11	12	13
11	2S	0.07090		
12 3 H 1S		-0.05173	0.21338	
13	2S	-0.02631	0.10963	0.07090

Full Mulliken population analysis:

		1	2	3	4	5
1 1 O 1S		2.06577				
2	2S	-0.01852	0.42318			
3	2PX	0.00000	0.00000	0.78905		
4	2PY	0.00000	0.00000	0.00000	0.52287	
5	2PZ	0.00000	0.00000	0.00000	0.00000	0.61747
6	3S	-0.05539	0.41771	0.00000	0.00000	0.00000
7	3PX	0.00000	0.00000	0.33099	0.00000	0.00000
8	3PY	0.00000	0.00000	0.00000	0.12797	0.00000
9	3PZ	0.00000	0.00000	0.00000	0.00000	0.22296
10 2 H 1S		-0.00117	0.01814	0.00000	0.05793	0.03207
11	2S	0.00208	-0.01310	0.00000	0.01776	0.00975
12 3 H 1S		-0.00117	0.01814	0.00000	0.05793	0.03207
13	2S	0.00208	-0.01310	0.00000	0.01776	0.00975

		6	7	8	9	10
6	3S	0.78660				
7	3PX	0.00000	0.54897			
8	3PY	0.00000	0.00000	0.12383		
9	3PZ	0.00000	0.00000	0.00000	0.32110	
10 2 H 1S		0.01586	0.00000	0.05973	0.04608	0.21338
11	2S	-0.05485	0.00000	0.02883	0.02391	0.07217
12 3 H 1S		0.01586	0.00000	0.05973	0.04608	-0.0285
13	2S	-0.05485	0.00000	0.02883	0.02391	-0.01166

		11	12	13
11	2S	0.07090		
12 3 H 1S		-0.01166	0.21338	
13	2S	-0.01362	0.07217	0.07090

Gross orbital populations:

1			
1 1 O 1S		1.99368
2	2S	0.83244
3	2PX	1.12004
4	2PY	0.80221
5	2PZ	0.92408
6	3S	1.07094
7	3PX	0.87996
8	3PY	0.42892
9	3PZ	0.68404
10 2 H 1S		0.49967
11	2S	0.13217
12 3 H 1S		0.49967

13   2S   0.13217
   Condensed to atoms (all electrons):
           1   2   3
1 O   8.250288 0.243017 0.243017
2 H   0.243017 0.428617 -0.039795
3 H   0.243017 -0.039795 0.428617
Total atomic charges:
          1
1 O   -0.736322
2 H   0.368161
3 H   0.368161
Sum of Mulliken charges = 0.00000
Atomic charges with hydrogens summed into heavy atoms:
          1
1 O   0.00000
2 H   0.00000
3 H   0.00000
Sum of Mulliken charges = 0.00000
Electronic spatial extent (au): $<R^{**}2> = 19.0349$
Charge = 0.00000 electrons
Dipole moment (Debye):
   X = 0.00000   Y = 0.00000   Z = -2.5271   Tot = 2.5271
Quadrupole moment (Debye-Ang):
   XX = -7.1299   YY = -4.2499   ZZ = -6.1015
   XY = 0.00000   XZ = 0.00000   YZ = 0.00000
Octapole moment (Debye-Ang = $^{**}2$):
   XXX = 0.00000   YYY = 0.00000   ZZZ = -1.3541   XYY = 0.00000
   XXY = 0.00000   XXZ = -0.3784   XZZ = 0.00000   YZZ = 0.00000
   YYZ = -1.1465   XYZ = 0.00000
Hexadecapole moment (Debye-Ang$^{**}3$):
   XXXX = -5.0259   YYYY = -5.9818   ZZZZ = -6.1729   XXXY = 0.00000
   XXXZ = 0.00000   YYYX = 0.00000   YYYZ = 0.00000   ZZZX = 0.00000
   ZZZY = 0.00000   XXYY = -2.0497   XXZZ = -1.9125   YYZZ = -1.7433
   XXYZ = 0.00000   YYXZ = 0.00000   ZZXY = 0.00000
N-N = 9.206554604692D+00 E-N = -1.990671058773D+02 KE = 7.597513288114D+01
Symmetry A1   KE = 6.773815321884D+01
Symmetry A2   KE = 0.000000000000D+00
Symmetry B1   KE = 4.624455014559D+00
Symmetry B2   KE = 3.612524647734D+00

11.6 CONCLUSION

Slater's 1951 paper proved to be a landmark in molecular structure theory (Slater 1951), and his theory was formally completed in 1965 by the work of Kohn and Sham (Kohn and Sham 1965). Slater's book 'The Calculation of Molecular Orbitals' (1979) tells the whole story.

12 Potential Energy Surfaces

I dealt briefly with potential energies and molecular geometries in earlier chapters. I also mentioned force constants. Well, the three concepts are inextricably linked so I will bring them together into a single chapter. People are interested in potential energy surfaces for a variety of reasons.

- They might want to calculate a molecular geometry, in which case they are interested in the lowest minimum point on the surface.
- They might want to fit an analytical expression to the potential energy surface for one reason or another, for example, to use the resulting curve as input to a scattering calculation, or to test out the validity of some physical model such as a harmonic oscillator. In the latter case, their physical model will be represented as a (simple) equation containing a number of parameters and the aim of the calculation is usually to find those parameters for comparison with experimental results.
- They might want to give a very accurate representation to the potential energy surface, with the intention of interpolating it, integrating it, differentiating it, etc. In this case, the form of the fitting equation is usually irrelevant.

12.1 A DIATOMIC MOLECULE

To get us going, consider a simple diatomic molecule like HCl. As usual, within the Born–Oppenheimer approximation, we focus attention on the electronic motions and calculate the electronic energy at a fixed internuclear distance. We then have to concern ourselves with the choice of 'level of theory'; in the case of HCl around the equilibrium bond length, it is probably adequate for most purposes to use HF/4–31G, so what I have done is to repeat the self consistent field calculation for a range of distances, giving the values in Table 12.1.

Table 12.1   HCl HF/4–31G Potential energy curve

R/Å	ϵ/E_h
1.20	−459.55756
1.25	−459.56231
1.30	−459.56362
1.35	−459.56234
1.40	−459.55910
1.45	−459.55438
1.50	−459.54855
1.55	−459.54188

Suppose that we are interested in testing the validity of the harmonic approximation, where

$$\epsilon = \epsilon_0 + \frac{1}{2}k_s(R - R_e)^2$$

and also as a matter of course, extracting the three values ϵ_0, k_s and R_e from this data. In order to find the dissociation energy, I would also have to run **atomic** HF/4–31G calculations on the atoms H(^{2}S) and Cl(^{2}S). I then used Jandel Scientific's 'Tablecurve2D' to give the best fit to this molecular data within the harmonic approximation, as shown in Figure 12.1.

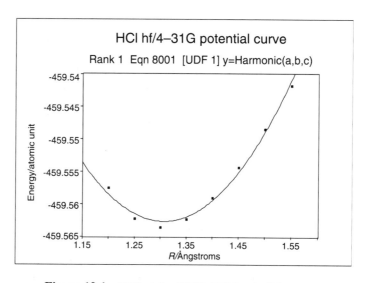

Figure 12.1   HCl at the HF/4–31G level of theory

Notice that the curve doesn't pass through all the points. Indeed, it doesn't necessarily have to pass through any of the points at all in order to give the best fit. The equation of the harmonic curve is

$$\epsilon/E_h = -459.56273 + 0.3671/(\text{pm})^2(R - 130.7\,\text{pm})^2$$

and the harmonic model parameters are shown in Table 12.2.

Table 12.2 Results for HCl.

level of fit	ϵ_0/E_h	$k/\text{N m}^{-1}$	R_e/pm
harmonic	−459.56273	320	130.7
best fit	−459.56362	455	129.8
experiment		516	127.4

The nuclei therefore vibrate in this potential, and in the harmonic approximation the vibrational energy levels are given by the standard formula

$$\epsilon_{\text{vib}} = hc_0\omega_e\left(v + \frac{1}{2}\right)$$

where ω_e is the fundamental vibrational wavenumber given by

$$\omega_e = \frac{1}{2\pi c_0}\sqrt{\frac{k_s}{\mu}}$$

When comparing the results of calculation with the results of thermochemical and/or spectroscopic measurements, it is necessary to remember the **zero point energy**. For example, the 'thermodynamic' dissociation energy D_e and the 'spectroscopic' disociation energies are related by

$$D_e = D_0 + \frac{1}{2}hc_0\omega_e$$

If, on the other hand, I had just wanted the best possible fit in order to interpolate the data, then my strategy would be different. Thus, the JANDEL curve fitting program tested several thousand different equations for goodness of fit and finally came up with an equation of the form

$$\epsilon = \epsilon_0 + a\ln(R') + b[\ln(R')]^2 + \ldots + f[\ln(R')]^6$$

where $R' = R/\text{Å}$. Obviously, the best fit equation need bear no relation to any simple physical model and, equally obviously, the equation is only valid for the range of the data used to calculate it.

In this case, I take the minimum of the curve as R_e, and the force constant is found by evaluating the second derivative at this value of R. Calculations of potential curves are expensive. If the aim of the study is to predict the geometry then it is not necessary to go to the expense of constructing the potential energy curve. The predicted bond length is the value of R at the minimum of the $\epsilon(R)$ curve, and so we need to mention ways of finding the minimum value of $\epsilon(R)$.

Minimization of an energy is not the only kind of minimizing problem encountered routinely in quantum chemistry, where we frequently have to be concerned with orbital localization, least-squares fitting of GTOs to STOs and so on. Such problems are not unique to theoretical chemistry, rather they belong to the branch of mathematics known as optimization theory. It is usual to divide minimization methods into two kinds:

- non-derivative methods and
- derivative methods,

and sometimes derivative methods are subdivided depending whether the first derivative, and/or the first and second derivatives are known. The literature is immense and a plethora of numerical methods exist.

12.1.1 A grid search

The simplest possible way of finding the minimum of $\epsilon(R)$ is to calculate the energy for a large number of different R values, and pick out the R corresponding to the lowest energy. This isn't a particularly good idea in standard *ab initio* calculations, as a significant resource has to be consumed in calculating each of the energy points. But in calculations such as molecular mechanics, where energy evaluation is particularly simple, the use of a grid search would be sensible. Usually, you choose a suitable interval for R in which you believe the minimum will lie, choose a distance at random within this interval, calculate the energy and repeat the calculation. The predicted bond length is then that corresponding to the lowest energy.

12.1.2 Derivatives and derivative methods

Consider a function $f(x)$ of a single variable, as shown in Figure 12.2.

In elementary calculus courses, we teach students about the first and second derivatives df/dx and d^2f/dx^2. At a minimum both $df/dx = 0$ and $d^2f/dx^2 > 0$. In such elementary courses, we often give a treatment of Newton's method for finding the zeros of such a function (the x values for which $f(x) = 0$).

The idea behind Newton's method is this. We start with some guess at the function zero, say $x^{(1)}$, and then fit a tangent line to the curve of the function at this point. The zero of this tangent line along the x axis then gives the next guess for the zero of the function, as shown on the figure. A simple calculation gives the point at which the tangent line cuts the x-axis as

$$x^{(2)} = x^{(1)} - \left[\frac{df(x^{(1)})}{dx}\right]^{-1} f(x^{(1)})$$

and this is taken as our next estimate of the root of the equation.

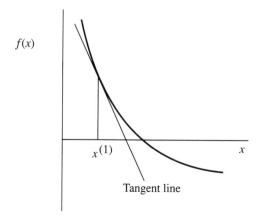

$f(x)$

$x^{(1)}$ x

Tangent line

Figure 12.2   Construct to demonstrate the Newton method for finding minima

Newton's method is usually formulated to give the roots of an equation $f(x) = 0$. The difference in optimization is that we want the **minimum** of $f(x)$, or a zero of the equation $df/dx = 0$. The method is therefore similar, except that the formula above for the next estimate of x becomes

$$x^{(2)} = x^{(1)} - \left[\frac{d^2 f(x^{(1)})}{dx^2}\right]^{-1} \frac{df(x^{(1)})}{dx}$$

and so we need to know the first and the second derivatives of $f(x)$ in order to apply this particular method.

Newton's method is derived from the Taylor series approximation for $f(x)$, which allows us to write the function value at a point $x + h$ in terms of the function value at point x as

$$f(x + h) = f(x) + \frac{h}{1!}\left(\frac{df(x)}{dx}\right) + \frac{h^2}{2!}\left(\frac{d^2 f(x)}{dx^2}\right) + \dots$$

and for small steps h we terminate the expansion at the second derivative, as shown. In this formula you have to calculate the derivative(s) and then evaluate them at the point of interest. To find the minimum, we seek the x value where $df/dx = 0$, and a little manipulation gives Newton's method.

The practicalities of the method depend on the ease of evaluating the function itself, together with the first and the second differential. Sometimes the function itself is easy to evaluate, sometimes the derivatives are impossible to evaluate as analytical expressions. In the case where analytical expressions are **not** readily available then the first and second order differentials have to be estimated numerically from finite differences. Finite difference calculations, themselves, involve evaluating the function at several points close to the point of interest. For example, a central differences two-point estimate gives

$$\frac{\mathrm{d}f(x)}{\mathrm{d}x} = \frac{f(x + \frac{1}{2}h) - f(x - \frac{1}{2}h)}{h}$$

for the first derivative of the function at point x. The calculation of this quantity involves two extra function evaluations. An estimate of the second derivative is given by the three point formula

$$\frac{\mathrm{d}^2 f(x)}{\mathrm{d}x^2} = \frac{f(x + h) + f(x - h) - 2f(x)}{2h}$$

We therefore have to consider the following

- Do we know the analytical first and second differentials?
- How easy is the function to evaluate?

In the case of a simple model like Molecular mechanics, the energy formula is very simple and direct differentiation is also easy. In standard *ab initio* calculations, the energy calculation is difficult since (e.g.) a HF calculation has to be performed. At first sight, differentiation of the HF energy expression with respect to an internuclear separation is easy, and many years ago a great deal of effort was expended on the following line of argument.

12.2   DIRECT DIFFERENTIATION: THE HELLMAN FEYNMAN THEOREM

Suppose we have a wavefunction $\Psi(c)$ that depends on a single parameter c. It doesn't matter what c is, it might be, for example, an orbital exponent or a bond distance and $\Psi(c)$ might be real or complex. Generally, we would be interested in calculating the variational integral which would depend on c

$$\epsilon(c) = \int \Psi^*(c) H \Psi(c) \mathrm{d}\tau$$

subject to the requirement that $\Psi(c)$ is correctly normalized.

$$\int \Psi^*(c) \Psi(c) \mathrm{d}\tau = 1$$

If I differentiate these two equations with respect to c, then we find

$$\frac{\mathrm{d}\epsilon(c)}{\mathrm{d}c} = \int \frac{\partial \Psi^*(c)}{\mathrm{d}c} H \Psi(c) \mathrm{d}\tau + \int \Psi^*(c) \frac{\partial H}{\mathrm{d}c} \Psi(c) \mathrm{d}\tau + \int \Psi^*(c) H \frac{\partial \psi(c)}{\mathrm{d}c} \mathrm{d}\tau$$

and

$$\int \frac{\partial \psi^*(c)}{\mathrm{d}c} \Psi(c)\mathrm{d}\tau + \int \Psi^*(c) \frac{\partial \Psi(c)}{\mathrm{d}c} \mathrm{d}\tau = 0$$

Normally, $\Psi(c)$ will be some approximation to the correct wavefunction. But if $\Psi(c)$ happens to be an eigenfunction of the Hamiltonian such that $H\Psi(c) = \epsilon\Psi(c)$, a little algebra shows that

$$\frac{\mathrm{d}\epsilon(c)}{\mathrm{d}c} = \int \Psi^*(c) \frac{\partial H}{\mathrm{d}c} \Psi(c) \, \mathrm{d}\tau$$

This result is referred to as the **Hellmann Feynman theorem**. It was widely used in the early days to investigate isoelectronic processes X $\rightarrow$ Y such as bond extensions, rotations of groups around molecular axes and so on. The idea was to write the left-hand side as $\Delta\epsilon(c)/\Delta c$, and that for an isoelectronic process the only non-zero contributions to the right-hand side would be due to changes in the nuclear positions and hence in the one-electron integrals.

Notice that the Hellmann Feynman theorem only applies to exact wavefunctions, not to variational approximations, and all the early optimism evaporated when it was realized eventually that such approximate wavefunctions themselves will also normally depend indirectly on the nuclear positions. For example, in an LCAO calculation, we fix the basis functions on the nuclear centres and when the nuclear centres move, the orbitals move with them. So the derivatives with respect to 'c' are not straightforward. I will have more to say about direct differentiation shortly, but let me mention a couple of problems with even simple potential energy surfaces.

12.3 MULTIPLE MINIMA

Polyatomic molecules are much more complicated than diatomics, and their potential energy surfaces show several complicating features. But even potential energy surfaces that depend on a single variable can also show interesting properties. Consider, for example, a model of ethane composed of two rigid CH_3 fragments which are joined through the C atoms but are free to rotate about the internuclear C-C axis. The potential energy surface is then a function of the single variable describing the azimuthal angle, and there are three identical minima at 60°, 180° and 300° (Figure 12.8). In this case the three minima are all equivalent.

Any search for a potential minimum would give one of the three equivalent minima. But suppose that we now substitute a fluorine for one of the H atoms in the 'left-hand' CH_3 group, and a Cl atom for one of the H atoms in the 'right hand' CH_3 group of ethane. For the sake of argument, we will keep the CH_2X fragments rigid, and only permit rotation about the C-C axis as before. The potential energy surface still depends on just the single variable, the azimuthal angle. But the surface is different and has two distinct types of minima. The azimuthal angle is set at 0° when the F and the Cl atoms are eclipsed. The minimum at 180° corresponds to the case where the F and Cl atoms are *trans* to each other. The other two minima, around 70° and 290°, correspond to intermediate cases. We say that each of the three minima is a **local minimum** and that the minimum at 180° is the **global minimum**.

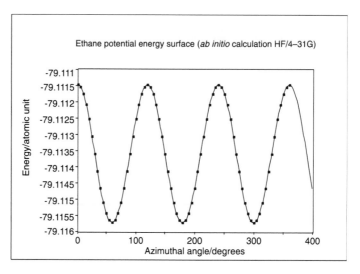

Figure 12.3   Potential energy for the rotation of a CH_3 group in ethane

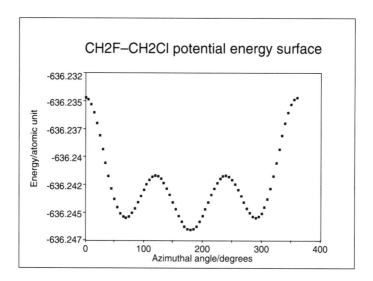

Figure 12.4   Potential energy surface for CH_2F-CH_2Cl about the C-C bond axis

Methods for finding minima normally can only find local minima. This is because we have to start at a given point on the x axis, and then make a next estimate by moving in one way or the other.

If we had started an energy minimization calculation at an azimuthal angle of $0°$, then we would almost certainly have ended up at the local energy minimum corresponding to an azimuthal angle of about $70°$. It may just happen that the local minimum we find is the true global minimum but, generally, we have no way of knowing whether this is true or not.

12.4   POTENTIAL ENERGY SURFACES THAT DEPEND ON SEVERAL VARIABLES

Consider the HF/4–31G potential energy calculation for H_2O shown in figure 12.5. Assuming C_{2v} symmetry, the energy depends on the common O-H bond length and the bond angle. What I did was to calculate the energy $\epsilon(R, \theta)$ for a grid of bond distances R and bond angles θ, and then fit the surface using Jandel's 'Tablecurve3D' package. The package chose the equation of best fit, which is given in the figure. As far as I am aware, this equation of best fit bears no relation to any physical model that has ever been tried for H_2O. It is simply what it says, the equation of best fit.

The minimum is given by the point $R = 95.1$ pm and $\theta = 111.2°$, and, in this particular case, there is a single minimum in the potential energy curve.

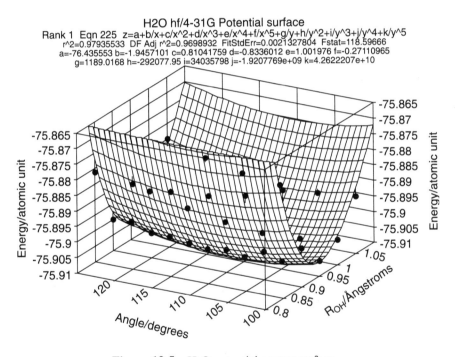

Figure 12.5 H_2O potential energy surface

12.4.1 Grid searches

The oldest method for finding the minimum is the **multivariate grid search method**, where we simply

- choose a suitable grid of points (e.g., 5 pm and 5° in the H_2O example discussed);
- select a suitable starting point (e.g. 100 pm and 100°);
- calculate the function value (the energy in this case) for the starting point and for all nearest neighbour surrounding points;
- choose the new approximation to the minimum and then iterate.

If the variables are independent, then this method will certainly be capable of reaching the minimum. If the variables are strongly dependent on one another, then this is not so and all kinds of practical problems can arise in the minimization.

The H_2O surface shown in Figure 12.5 is particularly simple because it shows only a single minimum in the range of variables for which I did the calculation. This minimum is also the global minimum.

A major problem when calculating potential energy surfaces for a polyatomic molecule is the problem of multiple minima, mentioned above. Many synthetic organic chemists, molecular biologists and biochemists are interested in large molecules containing hydrocarbon chains. We talk about **conformational energy searches** because the name of the game is to identify the global minimum and the low-energy conformations. Usually, we take the bond lengths as fixed in such calculations.

In the **Monte Carlo** method, which is widely used in MM methodology, we sample the possible conformations as follows. We

- choose a range of values for each of the geometric variables;
- choose at random a value for each variable within the allowed range;
- evaluate the energy.

The minimum is taken as the lowest energy value so obtained. Many MM packages collect automatically the possible conformations in this so-called 'window', arrange them in order of increasing energy and calculate their Boltzmann factors for a given temperature.

To give an idea of the magnitude of the problem, you might like to read an interesting benchmark paper by Saunders et al. These authors point out that there are 3^{14} possible conformations for the simple hydrocarbon cycloheptadecane.

Conformations of Cycloheptadecane. A Comparison of Methods
for Conformational Searching
Martin Saunders, K. N. Houk, Yun-Dong Wu, W. Clark Still, Mark Lipton,
George Chang and Wayne C. Guida
Journal of the American Chemical Society, **113** 1990 (112) 1419–1427

As a test of the effectiveness of various methods for searching the conformational space of highly flexible molecules, we conducted a series of conformational searches on cycloheptadecane. While none of the methods examined found all the low-energy conformations in a single search, all methods except distance geometry located the same global minimum. Some methods performed better than others at finding the other low-energy conformers. In all, 262 conformations of cycloheptadecane having MM2 energies within 3 kcal/mol of the global minimum were discovered. Among highly asymmetric structures, cycloheptadecane lies close to the boundary distinguishing problems which can and cannot be adequately addressed by using contemporary methodology and resources.

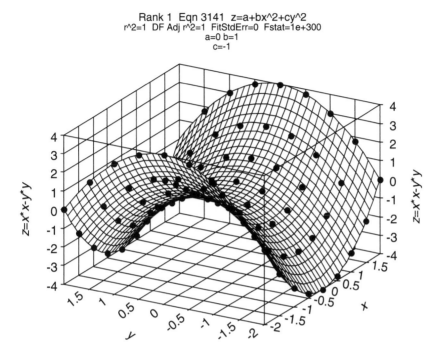

Figure 12.6   An example of a saddle point

12.4.2 Saddle points

Apart from minima, the other possibility of interest in chemistry is a saddle point. Saddle points don't occur for functions of a single variable; Figure 12.6, for illustration, shows that the function $f(x, y) = x^2 - y^2$ has a saddle point at the origin. You can see that the saddle point is a stationary point, and that the surface has a **minimum** with respect to x and a **maximum** with respect to y. At any stationary point, both $\partial f / \partial x$ and $\partial f / \partial y$ are zero.

For functions of two variables $f(x, y)$ such as that above, the elementary calculus texts rarely go beyond the simple observation that if the quantity

$$\frac{\partial^2 f}{\partial x^2} \frac{\partial^2 f}{\partial y^2} - \frac{\partial^2 f}{\partial x \partial y}$$

is greater than zero at a certain point, then we have either a minimum or a maximum, if the quantity is less than zero then there is neither a minimum or a maximum, and if the quantity equals zero then the case is undecided. A quick calculation for $f(x, y) = x^2 - y^2$ shows that this quantity is -4 at the origin (and indeed at all points in space). This simple criterion is not particularly useful for our study of potential energy surfaces.

12.4.3 The gradient and the hessian.

I mentioned the gradient of a function of three variables a little earlier in our very brief discussion of electromagnetic fields. For a function of three variables $f(x, y, z)$ we write

$$f = \frac{\partial f}{\partial x}\mathbf{e}_x + \frac{\partial f}{\partial y}\mathbf{e}_y + \frac{\partial f}{\partial z}\mathbf{e}_z$$

The gradient is a vector quantity, and it points in the direction of the maximum rate of **increase** of the function f. The negative -grad f, therefore points in the direction of the maximum rate of decrease. The gradient will of course change its direction from point to point, but if we could follow the direction of this negative gradient we should be led to the local minimum.

For a function of many variables $x_1, x_2, \ldots$ we generalize these concepts. The variables are written as a column vector $\mathbf{x} = (x_1, x_2, \ldots x_n)^T$ and the mathematical conditions which characterize the minimum of such a function $f(x_1, x_2, \ldots x_n)$ are expressed in terms of the **gradient vector g**

$$\mathbf{g} = \left(\frac{\partial f}{\partial x_1}, \frac{\partial f}{\partial x_2}, \ldots \frac{\partial f}{\partial x_n}\right)^T$$

and sometimes we need a knowledge of the second derivatives. These are collected into the matrix **g** called the **hessian**

$$\mathbf{G} = \begin{pmatrix} \dfrac{\partial^2 f}{\partial x_1{}^2} & \dfrac{\partial^2 f}{\partial x_1 \partial x_2} & \cdots & \dfrac{\partial^2 f}{\partial x_1 \partial x_n} \\[2ex] \dfrac{\partial^2 f}{\partial x_2 \partial x_1} & \dfrac{\partial^2 f}{\partial x_1{}^2} & \cdots & \dfrac{\partial^2 f}{\partial x_2 \partial x_n} \\[1ex] \cdots & & & \\[1ex] \dfrac{\partial^2 f}{\partial x_n \partial x_1} & \dfrac{\partial^2 f}{\partial x_n \partial x_2} & \cdots & \dfrac{\partial^2 f}{\partial x_n{}^2} \end{pmatrix}$$

A stationary point is one where the gradient is zero but the question then arises as to how we characterize such a stationary point. Normally, the only types of stationary point we are concerned with in chemistry are true minima, where the second derivatives are all positive, and transition states. Roughly speaking, a transition state corresponds to a special type of saddle point, with all the second derivatives positive except for one, which is negative. In order to investigate this behaviour, we calculate the eigenvalues and eigenvectors of the hessian matrix evaluated at the position of the stationary point. If all the eigenvalues of the hessian are positive, we have found a local minimum. If one eigenvalue is negative, we have found a transition state. The second derivatives are usually related to harmonic force constants, and very many packages will calculate these. I will give you an example shortly.

12.5 GOING DOWNHILL ON A POTENTIAL ENERGY CURVE

Many of the popular iterative methods are based on the simple principle of going downhill on a potential energy curve, such as that shown above for H_2O, in order to find the lowest energy point. The oldest of these methods is the method of steepest descents

12.5.1 Steepest descents

The negative gradient $-\mathbf{g}$, when evaluated at a point $\underline{x}$, points in the direction of 'steepest descent' down the surface away from $\underline{x}$, and the steepest descent method makes use of this idea. Actually, it isn't necessary to know the gradient vector exactly, as any rough numerical approximation will suffice. We calculate the gradient, and then choose a step size (how far to walk down the hill), improve our estimate of the minimum point accordingly and repeat until the minimum is found. The problem of the choice of step size is a very important one in most of these methods; it is usually estimated from a knowledge of previous steps. The steepest descents method has rather poor convergence properties near the minimum, and whilst it has been applied to the solution of the HF-LCAO equations, it has found little favour in studies of potential energy surfaces.

12.5.2 Second derivatives

Other methods require knowledge of the gradient **and** the second derivatives. As I mentioned above, the second derivatives of a function of many variables can be collected into the real symmetric hessian matrix $\underline{\mathbf{G}}$. The simple one-dimensional Newton method given above can be generalized quite transparently. We start from a 'point' $\underline{x}^{(1)}$ on the potential energy surface, and find a new point $\underline{x}^{(2)}$ given by

$$\underline{x}^{(2)} = \underline{x}^{(1)} - \underline{\mathbf{G}}^{-1}\mathbf{g}$$

where $\underline{\mathbf{G}}^{-1}$ is the inverse of the hessian matrix, calculated at the current point $\underline{x}^{(1)}$ and $\mathbf{g}$ is the gradient vector, also calculated at the point $\underline{x}^{(1)}$. We then repeat the process until we find the minimum.

The difficulty with this method is that we have to be able to calculate the function values (in this case the energy), together with the first derivatives and the second derivatives of the energy with respect to the geometric variables at each point in the search. The Newton method has found little popularity in molecular modelling. A number of methods get round the 'problem' of $\underline{g}^{-1}$ by 'updating' it at each iteration.

12.6 ANALYTICAL DIFFERENTIATION REVISITED

Over the last 20 years, a great deal of attention has been given to the development of analytical expressions for the first and second derivatives of the energy with respect to the molecular geometry, at the various levels of theory. Consider first a calculation at the HF-LCAO level of theory.

As I mentioned earlier, the energy depends directly on the nuclear coordinates $\mathbf{R}_A, \mathbf{R}_B, \ldots$ through the Hamiltonian. It depends indirectly on the values of R_α through the LCAO coefficients and also through the basis functions, since these are generally anchored on the nuclei and so move with them. It is usual to distinguish the **variational** from the **non-variational** parameters, because in the HF-LCAO procedure it turns out that the derivatives of the variational parameters (the LCAO coefficients) do not enter the gradient formula.

I explained earlier how the SCF equations could be written in terms of the various one- and two- electron integrals over basis functions in terms of the basis functions and the electron density matrix as

$$\epsilon_{\mathrm{el}} = \mathrm{trace}\,(\mathbf{h}_1 \mathbf{P}_1) + \frac{1}{2}\mathrm{trace}\,(\mathbf{P}_1 \mathbf{g})$$

where for example

$$G_{ij} = \sum \sum P_{1,kl}\left[\iint \chi_i(\mathbf{r}_1)\chi_j(\mathbf{r}_1)g(\mathbf{r}_1,\mathbf{r}_2)\chi_k(\mathbf{r}_2)\chi_l(\mathbf{r}_2)\mathrm{d}\tau_1\mathrm{d}\tau_2 \right.$$
$$\left. -\frac{1}{2}\iint \chi_i(\mathbf{r}_1)\chi_k(\mathbf{r}_1)g(\mathbf{r}_1,\mathbf{r}_2)\chi_j(\mathbf{r}_2)\chi_l(\mathbf{r}_2)\mathrm{d}\tau_1\mathrm{d}\tau_2 \right]$$

The total energy ϵ_{tot} is the sum of this electronic energy and the nuclear repulsion;

$$\epsilon_{\mathrm{tot}} = \epsilon_{\mathrm{el}} + \frac{e^2}{4\pi\epsilon_0}\sum_{\alpha=1}^{NUC-1}\sum_{\beta=\alpha+1}^{NUC}\frac{Z_\alpha Z_\beta}{R_{\alpha\beta}}$$

ϵ_{tot} depends on various geometric variables $x_1, x_2, \ldots x_p$, and we want to minimize it. The values of x are usually chosen to be the normal chemical variables: bond lengths and bond angles. It isn't necessary for you to be aware of the precise relationships between the cartesian coordinates and the so-called **valence force field**, but you are probably wondering about p in the list of variables.

For a molecule with NUC atoms, there are

- 3*NUC-6 vibrational degrees of freedom or
- 3*NUC-5 vibrational degrees of freedom if the molecule is linear.

So in principle, p cannot exceed this maximum. If you have chosen more than this number, then you must be dealing with dependent variables. Equally though, it may be that you only want to optimize the energy with respect to a few variables. Care and experience are needed in the choice of variables.

To calculate the gradient vector, it is necessary to differentiate the energy expression with respect to each of the x_α. The formula turns out to involve only $\underline{\mathbf{P}}_1$ and integrals involving derivatives of the basis functions with respect to the x_α, such as

$$\int \int \frac{\partial \chi_i(\mathbf{r}_1)}{\partial x_\alpha} \chi_j(\mathbf{r}_1) g(\mathbf{r}_1, \mathbf{r}_2) \chi_k(\mathbf{r}_2) \chi_l(\mathbf{r}_2) \, d\tau_1 \, d\tau_2$$

These can be easily evaluated when working with Gaussian orbitals, and the gradient calculated.

At other levels of theory, the distinction between the variational and non-variational parameters is not so useful. For example, in CI calculations, the energy expression does depend on the LCAO coefficients.

12.7 THE BERNY OPTIMIZATION ALGORITHM

Each major package tends to have its own geometry optimization package. The main geometry optimization procedure in GAUSSIAN92 is based on a program written by H. B. ('Berny') Schlegel, which implements his published algorithm (Schlegel 1982). The method is sophisticated, but at its core it uses the analytical gradient and a guessed hessian matrix. The inverse $\underline{\mathbf{G}}^{-1}$ is constantly updated during the optimization.

If analytical expressions for the gradients are not available at the particular level of theory chosen (or if the user so wishes), the package uses the Fletcher Powell method. This latter method uses information on the function to find a suitable interval for calculating each of the elements of the gradient vector by a simple forward-difference formula. If the error in this estimate is predicted to be too large, a more accurate central difference formula is used. The Fletcher Powell method is usually much slower than the Berny method.

12.8 A SIMPLE EXAMPLE

Back to H_2O, and I will **assume** C_{2v} symmetry. There are formally 3*3–6 vibrational degrees of freedom but the assumption of C_{2v} symmetry means that we only consider two of them for the geometry optimization. Users of GAUSSIAN92 generally choose to work with valence force fields so I have chosen the variables to be the bond length and the bond angle in my optimization. The input data is given in the box.

```
# HF/4-31G OPT

WATER MOLECULE

0 1

O
H1 O ROH
H2 O ROH H1 ANG

ROH 0.900
ANG   100.
```

There are three points to note. First the 'OPT' in the route. Then, I have implied C_{2v} symmetry by making each OH bond equal. If I had wanted to test whether the molecule really was of C_{2v} symmetry, I would have had to take independent OH distances. Finally I had to give an estimate of the variables in order to get the optimization started. The problem with the initial guess is that optimization methods generally tend to locate **local minima** and my initial guess could have been completely wrong. This is a major problem and the only known solution is to begin the optimization again from a different starting point.

Here is an extract from the first part of GAUSSIAN92 output for H_2O.

```
* * * * *******************************************************
Gaussian 92: HP-PARisc-HPUX-G92RevE.1 12-Jun-1993
   18-Oct-1994
* * * * *******************************************************


# HF/4-31G OPT


WATER MOLECULE

Symbolic Z-matrix:
   Charge = 0 Multiplicity = 1
O
H1  O  ROH
H2  O  ROH  H1  ANG
   Variables:
ROH        0.9
ANG        100.
GradGradGradGradGradGradGradGradGradGradGradGradGradGradGrad
Berny optimization.
Initialization pass.
                  -----------------------
                  ! Initial Parameters !
                  ! (Angstroms and Degrees) !
----------------               ------------------
! Name    Value   Derivative information (Atomic Units)   !
----------------------------------------------------------
! ROH      0.9          estimate D2E/DX2              !
! ANG      100.         estimate D2E/DX2              !
----------------------------------------------------------
Initial trust radius is 3.000D-01.
```

Points to note so far are the choice of the optimization procedure (the 'Berny' algorithm, which needs first and second derivatives), and the starting guess.

The program cycles for a while, with typical output

```
GradGradGradGradGradGradGradGradGradGradGradGradGradGradGrad
----------------------------------------------------------
         Z-MATRIX (ANGSTROMS AND DEGREES)
CD  Cent  Atom  N1  Length/X  N2  Alpha/Y  N3
----------------------------------------------------------
```

```
1  1   O
2  2   H    1    .954910( 1)
3  3   H    1    .954910( 2)    2   114.538( 3)
```

Z-Matrix orientation:

Center Number	Atomic Number	Coordinates (Angstroms) X	Y	Z
1	8	.000000	.000000	.000000
2	1	.000000	.000000	.954910
3	1	.868670	.000000	-.396568

Distance matrix (angstroms):

```
            1          2          3
1 O    .000000
2 H    .954910    .000000
3 H    .954910   1.606574    .000000
```
Interatomic angles:

H2-O1-H3 = 114.5378
STOICHIOMETRY H2O
FRAMEWORK GROUP C2V[C2(O),SGV(H2)]
DEG. OF FREEDOM 2
FULL POINT GROUP C2V NOP 4
LARGEST ABELIAN SUBGROUP C2V NOP 4
LARGEST CONCISE ABELIAN SUBGROUP C2 NOP 2
Standard orientation:

Center Number	Atomic Number	Coordinates (Angstroms) X	Y	Z
1	8	.000000	.000000	.103263
2	1	.000000	.803287	-.413053
3	1	.000000	-.803287	-.413053

Rotational constants (GHZ): 1059.0463813   388.5622227   284.2656604
Isotopes: O-16,H-1,H-1
Standard basis: 4-31G (S, S = P, 6D, 7F)
There are   7 symmetry adapted basis functions of A1 symmetry.
There are   0 symmetry adapted basis functions of A2 symmetry.
There are   2 symmetry adapted basis functions of B1 symmetry.
There are   4 symmetry adapted basis functions of B2 symmetry.
Crude estimate of integral set expansion from redundant
integrals = 1.247.
Integral buffers will be
262144 words long.
Raffenetti 1 integral format.
Two-electron integral symmetry is turned on.
   13 basis functions   28 primitive gaussians
    5 alpha electrons   5 beta electrons
       nuclear repulsion energy   9.1960146374 Hartrees.
One-electron integrals computed using PRISM.
One-electron integral symmetry used in STVInt
The smallest eigenvalue of the overlap matrix is 7.502D-02
DipDrv: MaxL = 4.
DipDrv: will hold 34 matrices at once.

Initial guess read from the read-write file:
Guess basis functions will be translated to current atomic
coordinates.
INITIAL GUESS ORBITAL SYMMETRIES.
   OCCUPIED (A1) (A1) (B2) (A1) (B1)
   VIRTUAL (A1) (B2) (B2) (B1) (A1) (A1) (B2) (A1)
alpha deviation from unit magnitude is 1.11D-15 for orbital 13.
alpha deviation from orthogonality is 5.55D-16 for orbitals 12 3.
A Direct SCF calculation will be performed.
Using DIIS extrapolation.
Closed shell SCF:
Requested convergence on RMS density matrix = 1.00D-08 within 64
cycles.
Requested convergence on MAX density matrix = 1.00D-06.
Two-electron integral symmetry used by symmetrizing Fock
matrices.
Keep R1 integrals in memory in canonical form, NReq =   433724.
IEnd =   28552 IEndB =   28552 NGot =   2048000 MDV =   2023763
LenX =   2023763
MinBra = 0   MaxBra = 3   MinLOS = -1 MaxLOS = -1   MinRaf = 0   MaxRaf = 3
MinLRy = 4.
IRaf = 0   NMat = 1   IRICut = 1   DoRegI = T   DoRafI = F
ISym2E = 1 JSym2E = 1.
Fock matrices symmetrized in FoFDir.
SCF DONE: E(RHF) = -75.9082698421   A.U. AFTER 10 CYCLES
   CONVG = .3182D-08 -V/T = 1.9998
   S**2 = .0000
KE = 7.592368118770D + 01   PE = -1.989091275677D + 02 EE =   3.788116190046D + 01
Compute integral first derivatives.
. . . and contract with generalized density number 0.
Use density number 0.
RysSet: KIntrp = 240 KCalc = 0 KAssym = 213
L702 exits . . . SP integral derivatives will be done elsewhere.
Compute integral first derivatives.
Integral derivatives from FoFDir, PRISM(SPDF).
MinBra = 0 MaxBra = 3 MinLOS = -1 MaxLOS = -1 MinRaf = 0 MaxRaf = 3
MinLRy = 4.
IRaf = 0 NMat = 1 IRICut = 1 DoRegI = T DoRafI = F
ISym2E = 1 JSym2E = 1.
Fock matrices symmetrized in FoFDir.
***** AXES RESTORED TO ORIGINAL SET *****

Center Number	Atomic Number	Forces (Hartrees/Bohr)		
		X	Y	Z
1	8	-.0018860	.0000000	-0.0012122
2	1	.0058227	.0000000	-.00698580
3	1	-.0039367	.0000000	.00819800

MAX .008197985 RMS .004351704

Internal Coordinate Forces (Hartree/Bohr or radian)

Cent	Atom	N1	Length/X	N2	Alpha/Y	N3	J

---

1 O
2 H   1   -.006986(1)

3 H   1   -.006986(2) 2   -.010507(3)

     MAX   .010507179 RMS   .008326714

> **Force constant calculations must be done at the same potential minimum, with the same level of theory**

GradGradGradGradGradGradGradGradGradGradGradGradGradGradGrad
Berny optimization.
Search for a local minimum.
Step number 2 out of a maximum of 20
All quantities printed in internal units
(Hartrees-Bohrs-Radians)
Update second derivatives using information from points 1 2
Trust test = 8.52D-01 RLast = 2.74D-01 DXMaxT set to 4.24D-01
The second derivative matrix:
                    ROH   ANG
     ROH   1.43670
     ANG   .04051 .21912
     Eigenvalues –   .21777 1.43805
RFO step: Lambda = -1.48473856D-04.
Quartic linear search produced a step of -.15041.

Variable	Old X	-DE/DX	Delta X (Linear)	Delta X (Quad)	Delta X (Total)	New X
ROH	1.80452	-.01397	-.01561	.00731	-.00829	1.79622
ANG	1.99906	-.01051	-.03816	-.01948	-.05764	1.94142

Item	Value	Threshold	Converged?
Maximum Force	.013972	.000450	NO
RMS Force	.012361	.000300	NO
Maximum Displacement	.057641	.001800	NO
RMS Displacement	.041178	.001200	NO

Predicted change in Energy = -4.327925D-04
GradGradGradGradGradGradGradGradGradGradGradGrad

> **and finally arrives at the local minimum**

Predicted change in Energy = -5.404687D-09
Optimization completed.
– Stationary point found.

                    Initial Parameters
                    (Angstroms and Degrees)
----------------   ------------------
Name	Value	Derivative information (Atomic Units)	
ROH	0.9505	-DE/DX =	-0.000122
ANG	111.2353	-DE/DX =	-0.000013

GradGradGradGradGradGradGradGradGradGradGradGradGradGradGradGrad

Notice that the gradient is indeed zero at this point, but these algorithms usually find a stationary point, which might be a local minimum; it might be a saddle point and so on. Even if it is a local minimum, we have no way of knowing whether the point found is the global minimum.

```
# HF/4-31G FREQ

WATER MOLECULE

0  1

O
H1  O  ROH
H2  O  ROH  H1  ANG

ROH   0.9505
ANG   111.2353
```

12.9   FORCE CONSTANTS

The next step in all these optimizations should be to investigate the nature of the stationary point and this can be done easily by the calculation of harmonic force constants. In most *Ab Initio* packages, all that is necessary is to specify the molecular geometry **at the stationary point** and changing the route card. The geometry **MUST** correspond to the stationary point geometry because we need to calculate quantities like the second derivative **at the position of the stationary point**. There is a bonus because the calculated harmonic vibrational frequencies can be compared with experiment. For the same reason as above, the 'level of theory' has to be identical. Force constants are usually calculated within the harmonic approximation. In the GAUSSIAN suite, it is only necessary to make the changes shown in the box.

Output from the GAUSSIAN92 package is standard

```
----------------
# HF/4-31G FREQ
----------------
----------------
WATER MOLECULE
----------------
Symbolic Z-matrix:
   Charge = 0 Multiplicity = 1
O
H1 O ROH
H2 O ROH H1 ANG
   Variables:
ROH 0.9505
ANG 111.2353
GradGradGradGradGradGradGradGradGradGradGradGradGrad
Berny optimization.
Initialization pass.
```

```
------------------------
! Initial Parameters !
! (Angstroms and Degrees) !
```
```
----------------                    ------------------
! Name      Value   Derivative information (Atomic Units)   !
```
```
-----------------------------------------------------------------
! ROH       0.9505     calculate D2E/DX2 analytically !
! ANG     111.2353     calculate D2E/DX2 analytically !
-----------------------------------------------------------------
```
until we get down to the force constant calculations
Compute integral second derivatives.
Integral derivatives from FoFDir, PRISM(SPD) Scalar Rys(F).
MinBra = 0 MaxBra = 2 MinLOS = -1 MaxLOS = -1 MinRaf = 0 MaxRaf = 2 MinLRy = 3.
IRaf = 0 NMat = 1 IRICut = 1 DoRegI = T DoRafI = F ISym2E = 1 JSym2E = 1.
Fock matrices symmetrized in FoFDir.
Full mass-weighted force constant matrix:
Low frequencies —   —.0022   .0017   .0025   22.1912   24.0860   24.8313
Low frequencies —   1742.8979
3957.6864
4109.4644
Harmonic frequencies (cm**-1), IR intensities (KM/Mole),
Raman scattering activities (A**4/AMU), Raman depolarization ratios,
reduced masses (AMU), force constants (mDyne/A) and normal coordinates:

			1	2	3
A1			A1	B2	
Frequencies —			1742.8979	3957.6864	4109.4644
Reduced masses —			1.0917	1.0369	1.0884
Force constants —			1.9538	9.5694	10.8299
IR Intensities —			119.7002	2.7688	49.7417
Raman Activities —			10.5397	90.0170	40.2162
Depolarizations —			.3947	.2196	.7500
Coord	Atom	Element:			
1	1	8	.00000	.00000	.00000
2	1	8	.00000	.00000	-.07334
3	1	8	-.07479	-.04407	.00000
1	2	1	.00000	.00000	.00000
2	2	1	.38079	-.61377	.58199
3	2	1	.59347	.34974	-.39824
1	3	1	.00000	.00000	.00000
2	3	1	-.38079	.61377	.58199
3	3	1	.59347	.34974	.39824
```
Harmonic frequencies (cm**-1), IR intensities (KM/Mole),
Raman scattering activities (A**4/AMU), Raman depolarization ratios,
reduced masses (AMU), force constants (mDyne/A) and normal coordinates:

| | | | | 1 | | | 2 | | | 3 | |
|---|---|---|---|---|---|---|---|---|---|---|---|
| | | | | A1 | | | A1 | | | B2 | |
| Frequencies – | | | | 1742.8979 | | | 3957.6864 | | | 4109.4644 | |
| Red. masses – | | | | 1.0917 | | | 1.0369 | | | 1.0884 | |
| Frc consts – | | | | 1.9538 | | | 9.5694 | | | 10.8299 | |
| IR Inten – | | | | 119.7002 | | | 2.7688 | | | 49.7417 | |
| Raman Activ – | | | | 10.5397 | | | 90.0170 | | | 40.2162 | |
| Depolar – | | | | .3947 | | | .2196 | | | .7500 | |
| Atom | AN | X | Y | Z | X | Y | Z | X | Y | Z | |
| 1 | 8 | .00 | .00 | -.07 | .00 | .00 | -.04 | .00 | -.07 | .00 | |
| 2 | 1 | .00 | .38 | .59 | .00 | -.61 | .35 | .00 | .58 | -.40 | |
| 3 | 1 | .00 | -.38 | .59 | .00 | .61 | .35 | .00 | .58 | .40 | |
```

and the results should be self explanatory.

All three vibrational modes are infrared active, and the results are compared with experimental results in Table 12.3.

Table 12.3 H_2O vibrational analysis, HF/4–31G level of theory

Mode	Symmetry	ω_e/cm^{-1} Experiment	Calculated
OH symmetric stretch	A_1	3652	3958
OH asymmetric stretch	B_2	3756	4109
HOH bend	A_1	1595	1742

For polyatomic molecules, SCF calculations generally overestimate the force constants and are usually scaled in order to give reliable predictions. Many of these *ab initio* packages also give information that could be useful for comparing experimental and theoretical estimates of thermodynamic data (remember the zero point energies!).

12.10 THERMOCHEMICAL CALCULATIONS

```
----------------
- THERMOCHEMISTRY -
----------------
TEMPERATURE 298.150 KELVIN.
PRESSURE 1.00000 ATM.
ATOM 1 HAS ATOMIC NUMBER 8 AND MASS   15.99491
ATOM 2 HAS ATOMIC NUMBER 1 AND MASS   1.00783
ATOM 3 HAS ATOMIC NUMBER 1 AND MASS   1.00783
Molecular mass:   18.01056 amu.
Principle axes and moments of inertia in atomic units:
                    1         2         3
EIGENVALUES –    1.84173   4.42922   6.27096
X                  .00000    .00000  -1.00000
Y                -1.00000    .00000    .00000
Z                  .00000   1.00000    .00000
THIS MOLECULE IS AN ASYMMETRIC TOP.
ROTATIONAL SYMMETRY NUMBER 2.
ROTATIONAL TEMPERATURES (KELVIN)   47.02819  19.55499  13.81183
ROTATIONAL CONSTANTS (GHZ)   979.91411   407.46221   287.79356
ZERO-POINT VIBRATIONAL ENERGY   58677.1 (JOULES/MOL)
     14.02417 (KCAL/MOL)
     .0223489 (HARTREE/PARTICLE)
VIBRATIONAL TEMPERATURES:   2507.63   5694.20   5912.57
     (KELVIN)
SUM OF THERMAL ENERGIES:   .0251833 (HARTREE/PARTICLE)
SUM OF HARTREE-FOCK AND THERMAL ENERGIES:   -75.8834526
(HARTREE/PARTICLE)
```

	E	CV	S
	JOULES/MOL	JOULES/MOL-KELVIN	JOULES/MOL-KELVIN
TOTAL	66118.684	25.074	188.062
ELECTRONIC	.000	.000	.000

TRANSLATIONAL	3718.457	12.472	144.802
ROTATIONAL	3718.457	12.472	43.243
VIBRATIONAL	58681.770	.131	.017
	E	CV	S
	KCAL/MOL	CAL/MOL-KELVIN	CAL/MOL-KELVIN
TOTAL	15.803	5.993	44.948
ELECTRONIC	.000	.000	.000
TRANSLATIONAL	.889	2.981	34.609
ROTATIONAL	.889	2.981	10.335
VIBRATIONAL	14.025	.031	.004
	Q	LOG10(Q)	LN(Q)
TOTAL BOT	.638827E-02	-2.194617	-5.053514
TOTAL V=0	.121650E+09	8.085112	18.616435
VIB (BOT)	.525253E-10	-10.279632	-23.669727
VIB (V=0)	.100022E+01	.000097	.000223
ELECTRONIC	.100000E+01	.000000	.000000
TRANSLATIONAL	.300436E+07	6.477751	14.915574
ROTATIONAL	.404821E+02	1.607264	3.700861

***** AXES RESTORED TO ORIGINAL SET *****

Center Number	Atomic Number	Forces (Hartrees/Bohr) X	Y	Z
1	8	.000027925	.000000000	.000019108
2	1	.000005723	.000000000	-.000038323
3	1	-.000033648	.000000000	.000019215

MAX .000038323 RMS .000021468

Internal Cent	Atom	N1	Coordinate Length/X	N2	Alpha/Y	N3	Beta/Z	J
1	O							
2	H	1	-.000038(1)					
3	H	1	-.000038(2)	2	-.000010(3)			

MAX .000038323 RMS .000031849

FORCE CONSTANTS IN CARTESIAN COORDINATES (HARTREES/BOHR).

	1	2	3	4	5
1	.553804D+00				
2	.000000D+00	.333598D-04			
3	-.173960D+00	.000000D+00	.689000D+00		
4	-.544232D-01	.000000D+00	-.372283D-01	.550959D-01	
5	.000000D+00	-.166799D-04	.000000D+00	.000000D+00	.200983D-04
6	.382856D-01	.000000D+00	-.566978D+00	-.221195D-01	.000000D+00
7	-.499380D+00	.000000D+00	.211189D+00	-.672629D-03	.000000D+00
8	.000000D+00	-.166799D-04	.000000D+00	.000000D+00	-.341837D-05
9	.135675D+00	.000000D+00	-.122021D+00	.593478D-01	.000000D+00

	6	7	8	9
6	.584431D+00			
7	-.161661D-01	.500053D+00		
8	.000000D+00	.000000D+00	.200983D-04	
9	-.174523D-01	-.195022D+00	.000000D+00	.139474D+00

FORCE CONSTANTS IN INTERNAL COORDINATES (ATOMIC UNITS).

	1	2	3
1	.584431D + 00		
2	-.874719D-02	.584431D + 00	
3	.397364D-01	.397364D-01	.177686D + 00

GradGradGradGradGradGradGradGradGradGradGradGradGradGradGradGradGradGrad
Berny optimization.

We were lucky, because the stationary point on the H_2O surface turned out to be a minimum point. When this happens, we get the 'correct' number (3*NUC-6) of vibrational frequencies. For more complicated molecules, the stationary point can often turn out to be a saddle point and in this case we get negative frequencies. A saddle point that is a maximum in one direction is known to chemists as a **transition state**. It is characterized by a single negative frequency on the potential energy surface.

12.11 A TRANSITION STATE

It would be wrong to close this chapter without giving you a real example of a transition state. I have just chosen one particular example, from a recent journal.

Theoretical Analysis of the Thermal and Lewis Acid Catalyzed Reactions of Butadiene and Acrolein.

M. I. Menédez, J. González, J. A. Sordo and T. L. Sordo
Journal of Molecular Structure (THEOCHEM), **314** (1994) 241

The thermal and Lewis acid catalyzed reactions of butadiene and acrolein have been analyzed theoretically. It is found that one of the new C-C bonds (C1-C6) is formed mainly through charge transfer from acrolein to butadiene. Owing to the electrophilicity of the dienophile, the HOMOI-LUMO monotransference from butadiene to acrolein becomes predominant over that from acrolein to butadiene. As a consequence, transition structures present asynchronicity and a diminution of the activation energy is produced. These effects are more accentuated for the *s-cis* transition structures given the greater electrophilicity of the *s-cis* conformation of acrolein. The endo preference is characterized by the lesser weight of the costlier transference from the HOMO of butadiene to the NLUMO of acrolien which lead directly to no bonding. The catalytic effect of BH_3 reinforces all these effects.

The Diels–Alder reaction is the most general synthetic method for obtaining six-membered rings. Recently, *Ab Initio* calculations on the reaction of butadiene with acrolein have given us useful data on stereochemical effects on activation energies, and on the stereochemistry of the process. The chemical reaction is shown in Figure 12.7.

The Diels–Alder reaction and in this particular paper the authors chose R = H, R = CHO and R = CHO-BH_3. They performed HF/3–21G OPT level calculations, and located the transition structure using the Schlegel algorithm. In each case, they checked (correctly) that the hessian matrix only had one negative eigenvalue.

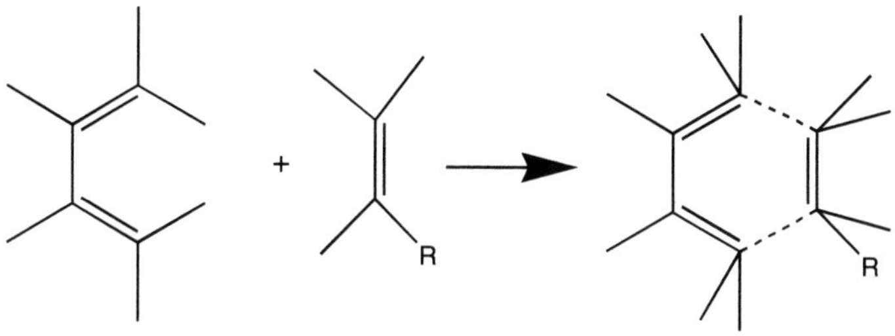

Figure 12.7   The Diels–Alder reaction

13 Primary Properties

In their classic paper 'Mathematical Problems in the Complete Quantum Predictions of Chemical Phenomena', Boys and Cook (1962) divide the determination of an *Ab Initio* electronic wavefunction into distinct logical stages, which include the

- choice of atomic orbital basis set;
- calculation of one-electron integrals;
- calculation of two-electron integrals;
- choice of Slater determinants, and finally the;
- calculation of the wavefunction.

Once the wavefunction has been determined, a number of molecular properties X can be determined from expressions such as

$$X_{el} = \int \Psi_{el} \sum X(\mathbf{r}_i) \Psi_{el} d\tau$$

where X_{el} is the contribution made by the electrons to the property X (for example, the electric dipole moment), and the sum is over all the electrons. Within the context of the Born–Oppenheimer approximation, we then have to remember to add a corresponding contribution from the nuclei.

Boys and Cook refer to these properties as **primary properties** because their electronic contributions can be obtained directly from the electronic wavefunction Ψ_{el}. As a matter of interest, they also classified the electronic energy as a primary property. I decided to treat this quantity in a separate chapter 12 in view of its importance. In any case, it cannot be calculated as the expectation value of a sum of true one-electron operators. The one-electron operators of SCF theory include the average effects of the other electrons, and are sometimes called 'pseudo' one-electron operators for this reason.

Several other physical properties are given by the rate of change of the primary property with respect to a change in the positions of the nuclei and so on. For instance, the rate of change of the energy with respect to small variations in the nuclear positions gives the force constants and the rate of change of the electric dipole moment with nuclear positions determines the intensities of the bands in vibrational infrared spectra. These properties are referred to by Boys and Cook as **derivative properties**. **Induced properties** are those that are induced by the application of some external field. I am going to concentrate on polarizabilities and magnetizabilities in a later chapter 14. Such properties measure the response of a molecule to the applied field, and are also referred to as response functions. Finally, Boys and Cook refer to **interactions between systems** as the fourth category of molecular property. This is not a unique way of classifying properties. For example, Dykstra et al (1990) show how a property such as the electric dipole moment can be defined in terms of the first derivatives of the energy with respect to an applied electric field. I am going to stick with the Boys and Cook's nomenclature as a framework for discussion.

Much of our knowledge of molecules is obtained from experimental studies of the way they interact with electromagnetic radiation, and the recent rapid growth in non-linear spectroscopy and molecular electronics has focused attention on our ability (or otherwise) to predict and rationalize the electric properties of molecules. The idea of an electric multipole is an important one, so let's begin there.

13.1 ELECTRIC MULTIPOLE MOMENTS

Can I refer you to Figure 13.1, which is similar to that in Chapter 0 (where I was concerned with the forces between point charges. In the figure, we see a pair of point charges; Q_A at position vector $\underline{r}_A$ and Q_B at position vector $\mathbf{r}_B$. Their *electric dipole moment* $\underline{\mathbf{p}}_e$ is defined as

$$\underline{\mathbf{p}}_e = Q_A\underline{r}_A + Q_B\underline{r}_B$$

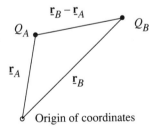

Figure 13.1   Construct needed to define a simple electric dipole

which is consistent with the more elementary definition for a pair of equal and opposite charges $\pm Q$ separated by distance d (in which case the dipole moment has magnitude Qd and direction from $-Q$ to $+Q$). Suppose for the sake of generality that point charges $Q_A, Q_B, \ldots, Q_N$ are located at positions $\underline{r}_A, \underline{r}_B \ldots \underline{r}_N$ respectively. The general definition of the electric dipole moment for this charge distribution is

$$\underline{\mathbf{p}}_e = \sum_{i=A}^{N} Q_i \underline{\mathbf{r}}_i$$

The vector $\underline{\mathbf{p}}_e$ thus has three components whose values are $\sum Q_i x_i \sum Q_i y_i$ and $\sum Q_i z_i$. This definition of the electric dipole moment is independent of the choice of coordinate origin, provided that the charges sum to zero. But if you intend reporting the dipole moment of an ion, it is **obligatory** that you specify the coordinate origin.

At the molecular level, electric dipole moments are important because they give information about the charge distribution in a molecule. Examination of the experimental data for a few simple compounds reveals that the electric dipole moment is also a property associated with chemical bonds and their polarity. The literature is immense, and you might like to consult the texts by Smith or LeFèvre on this subject.

The quantities $\sum Q_i x_i^2, \sum Q_i y_i^2, \sum Q_i z_i^2, \sum Q_i x_i y_i, \sum Q_i x_i z_i$ and $\sum Q_i y_i z_i$ define the **electric second moments** which we can write as a 3×3 symmetric matrix

$$\begin{pmatrix} \sum Q_i x_i^2 & \sum Q_i x_i y_i & \sum Q_i x_i z_i \\ \sum Q_i y_i x_i & \sum Q_i y_i^2 & \sum Q_i y_i z_i \\ \sum Q_i z_i x_i & \sum Q_i z_i y_i & \sum Q_i z_i^2 \end{pmatrix}$$

Unfortunately, there are several different definitions relating to second moments in the literature, and you will have to be careful when applying seemingly simple formulae that involve such quantities. Not only that, but people use the same name for the different definitions. There is actually a reason for this, not only connected with perversity. Just trust me for now.

Most authors prefer to work with a quantity Θ called the **electric quadrupole moment**, defined as

$$\underline{\Theta} = \frac{1}{2} \begin{pmatrix} \sum Q_i(3x_i^2 - r_i^2) & 3\sum Q_i x_i y_i & 3\sum Q_i x_i z_i \\ 3\sum Q_i y_i x_i & \sum Q_i(3y_i^2 - r_i^2) & 3\sum Q_i y_i z_i \\ 3\sum Q_i z_i x_i & 3\sum Q_i z_i y_i & \sum Q_i(3z_i^2 - r_i^2) \end{pmatrix}$$

The factor of ½ is conventional. Adding together the diagonal elements of Θ gives a sum of zero, by definition. The diagonal elements of Θ always add to zero, for any charge distribution. For a spherically symmetrical charge distribution, each of the diagonal terms is zero, and so the electric quadrupole moment gives a measure of the deviation of a charge distribution from spherical symmetry.

At the molecular level, quadrupoles can lead to useful structural information. Thus, whilst the absence of a dipole moment in CO_2 simply means that the molecule is linear, the fact that the electric quadrupole moment is negative shows that our simple chemical intuition of $O^-C^+O^-$ is correct. This definition of the electric quadrupole moment is independent of the choice of coordinate origin only when the charges sum to zero **and** the electric dipole moment is zero. So if you intend reporting a quadrupole moment you will usually have to specify the coordinate origin.

Depending on the orientation of the x, y and z axes we choose for a given array of point charges, the matrix of quadrupole moments may or may not be diagonal

$$\underline{\Theta} = \begin{pmatrix} \Theta_{a,a} & 0 & 0 \\ 0 & \Theta_{b,b} & 0 \\ 0 & 0 & \Theta_{c,c} \end{pmatrix}$$

If for some particular choice of axis system the matrix **is** diagonal, we refer to the non-zero components as the **principal values** of this property, and we talk about the principal axes. The largest of these three components is usually referred to as 'the' quadrupole moment. In any case, $\Theta_{aa} + \Theta_{bb} + \Theta_{cc} = 0$. In order to determine the principal values and the principal axes, we start from a 'non-diagonal' Θ and find the eigenvalues and eigenvectors. The eigenvalues of Θ are the principal values, and the eigenvectors determine the directions of the principal axes.

In chemistry we usually need to consider continuous distributions of electric charge $\rho(\underline{r})$ such as that due to the electrons, in addition to the array of nuclear point charges which are fixed in space according to the Born–Oppenheimer approximation. As far as continuous charge distributions are concerned, the discussion above is unchanged except that we replace sums such as $\sum Q_i x_i^2$ and $\sum Q_i y_i z_i$ by corresponding integrals over the charge distribution $\int \rho(\underline{r}) x^2 d\tau$ and $\int \rho(\underline{r}) yz d\tau$. Molecular electronic contributions to the second moments are therefore given by

$$\begin{pmatrix} \int \rho(\underline{r}) x^2 d\tau & \int \rho(\underline{r}) xy d\tau & \int \rho(\underline{r}) x_i z_i d\tau \\ \int \rho(\underline{r}) yx d\tau & \int \rho(\underline{r}) y^2 d\tau & \int \rho(\underline{r}) yz d\tau \\ \int \rho(\underline{r}) zx d\tau & \int \rho(\underline{r}) zy d\tau & \int \rho(\underline{r}) z^2 d\tau \end{pmatrix}$$

The physical idea behind electric multipole moments is that they describe collectively a charge distribution; a charge density can be regarded as being made up from an overall charge, an electric dipole, an electric quadrupole, and so on. Look again at my discussion of electrostatics in Chapter 0, where I gave you Figure 13.2 in order to discuss the electrostatic field at an arbitrary point in space. Suppose now that the

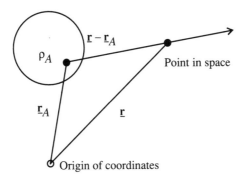

Figure 13.2   Construct needed to define the electric field at a point in space, due to a charge distribution

'charge distribution' ρ_A corresponds to an atom or a molecule. In principle, we should be able to calculate the electrostatic potential $\varphi(\underline{r})$ at the position $\underline{r}$ by using the basic definition

$$\varphi(\underline{r}) = \frac{1}{4\pi\epsilon_0} \int \frac{\rho(\underline{r}_A)}{|\underline{r} - \underline{r}_A|} d\tau_A$$

to add contributions from all points in the charge distribution. But as you might anticipate, in practice there are usually immense problems in evaluating the integral, depending on the symmetry of the charge distribution.

The multipole expansion gives a method of calculating the potential in terms of the potentials due to the charge, the electric dipole, the electric quadrupole and the higher poles associated with the charge distribution. Algebraic expressions for these terms are also much easier to deduce; for example, the potential due to the dipole $\underline{p}_e$ is shown in all the elementary electromagnetism texts to be

$$\varphi_{\text{dipole}}(\underline{r}) = \frac{\underline{p}_e \cdot \underline{r}}{4\pi\epsilon_0 r^3}$$

13.1.1 The multipole expansion

If the charge distribution shown in Figure 13.2 has overall charge Q, electric dipole moment $\underline{p}_e$, electric quadrupole moment $\underline{\Theta}_e$ and so on then the potential at $\underline{r}$ can be written as a sum of contributions due to the overall charge, the electric dipole, the electric quadrupole, etc. and this forms the so-called **multipole expansion**.

$$\varphi(\underline{r}) = \varphi_{\text{charge}}(\underline{r}) + \varphi_{\text{dipole}}(\underline{r}) + \varphi_{\text{quadrupole}}(\underline{r}) + \cdots$$

The potential due to the overall charge falls off as quickly as $1/r$, the potential due to the electric dipole falls off as $1/r^2$, the potential due to the electric quadrupole falls off as $1/r^3$ and so on. So we would hope that only the first few terms in the expression would be needed, and this explains why you rarely hear about moments higher than the second.

It is conventional to write the multipole expansion as a series in $1/r$, and so we need to find alternative expressions for the dipole and higher order terms. From elementary vector calculus we have that

$$\text{grad}\,(1/r) = -\frac{\underline{r}}{r^3}$$

So we write

$$(4\pi\epsilon_0)\phi(\underline{r}) = \frac{Q}{r} - \underline{p}_e \cdot \text{grad}\left(\frac{1}{r}\right) + \cdots$$

where each term on the right hand side is a product of an electrical moment with an expression involving $1/r$.

For a long time, the multipole expansion was regarded widely as giving the correct starting point for a study of intermolecular forces. As I mentioned above, the hope is that the major contributions to the potential around a molecule will be those associated with the overall charge (if any), the dipole and perhaps the quadrupole, and that contributions from higher electric multipoles will be negligible. It is interesting to note that all the classic texts on intermolecular forces begin with a discussion of electromagnetism and the multipole expansion. Historically, such studies were also an indirect route to molecular quadrupole moments.

It is often useful to know the energy of interaction U of a charge distribution when it is placed in an external electrostatic field

$$U = Q\varphi - \underline{\mathbf{p}}_e.\underline{\mathbf{E}} - \frac{1}{3}\sum\sum \Theta_{ij}\underline{E}'_{ij} - \cdots$$

where ϕ is the electrostatic potential and $\underline{\mathbf{E}}'$ the electric field gradient. The field gradient is defined by the following 3×3 matrix

$$\underline{\mathbf{E}}' = \begin{pmatrix} \dfrac{\partial E_x}{\partial x} & \dfrac{\partial E_x}{\partial y} & \dfrac{\partial E_x}{\partial z} \\[2ex] \dfrac{\partial E_y}{\partial x} & \dfrac{\partial E_y}{\partial y} & \dfrac{\partial E_y}{\partial z} \\[2ex] \dfrac{\partial E_z}{\partial x} & \dfrac{\partial E_z}{\partial y} & \dfrac{\partial E_z}{\partial z} \end{pmatrix}$$

Just like the electric quadrupole discussed above, the electric field gradient will sometimes have diagonal form, depending on the choice of axes used. We speak again of the principal values, and the principal axes, and determine them in the same way.

13.1.2 Calculations of electric moments

Electric dipole moments can be determined experimentally from the Stark effect, or from a study of dielectric polarization. The Stark effect gives the **modulus** of the electric dipole, which of course is a vector quantity. The **direction** of the vector in the case of a linear molecule can be deduced from an isotope substitution study of the rotational molecular **magnetic** moment.

The electric dipole moment operator is

$$\underline{\mathbf{p}}_e = e\sum_{\alpha=A}^{NUC} Z_\alpha\underline{\mathbf{R}}_\alpha - e\sum_{i=1}^{n}\underline{\mathbf{r}}_i$$

where the first sum runs over the NUC nuclei and the second over the n electrons. Z_α is the atomic number of nucleus α. Within the Born–Oppenheimer approximation, the nuclei are to be thought of as being clamped in space and we, therefore, calculate the electric dipole moment according to

$$\underline{\mathbf{p}}_e = e \sum_{\alpha=A}^{NUC} Z_\alpha \underline{\mathbf{R}}_\alpha - e \int \psi_{el} \sum_{i=1}^{n} \mathbf{r}_i \psi_{el} d\tau$$

which reminds us the property has a nuclear contribution and an electronic contribution.

The electronic wavefunction depends on the space and spin coordinates of all the n electrons present. The dipole moment operator is made up from a sum of one-electron spinless operators. In previous chapters, I have emphasized the importance of the electron density $P_1(\mathbf{r})$ and explained how it can be determined from a many-electron wavefunction like Ψ_{el} above. $P_1(\mathbf{r})$ is a number density, it gives the number of electrons per unit volume. The charge density is $-eP_1(\mathbf{r})$ and we rewrite the dipole moment expression above as

$$\underline{\mathbf{p}}_e = e \sum_{\alpha=A}^{NUC} Z_\alpha \underline{\mathbf{R}}_\alpha - e \int P_1(\mathbf{r})\mathbf{r} d\tau$$

where the variable $\mathbf{r}$ relates to an arbitrary electron.

Usually we use a basis set $\chi_1, \chi_2 \ldots \chi_n$, and again I told you earlier that we represent $P_1(\mathbf{r})$ as an $n \times n$ matrix. We also collect together the dipole integrals such as $\int \chi_1(\mathbf{r}) x \chi_1(\mathbf{r}) d\tau$ as an $n \times n$ matrix.

The technology is simple: we

- calculate the electronic wavefunction;
- calculate the dipole integrals over the basis functions χ_i;
- calculate the dipole moment.

Exactly the same comments go for the higher electric moments, and they are often output routinely following an SCF calculation. For example, our H_2O STO/4G calculation using GAUSSIAN92 gives the output shown below.

```
Charge =    .0000 electrons
Dipole moment (Debye):
     X =   .0000      Y =     .0000      Z =  -1.7634    Tot =   1.7634
Quadrupole moment (Debye-Ang):
    XX =  -6.0801     YY =   -4.3053     ZZ =  -5.3664
    XY =   .0000      XZ =     .0000     YZ =    .0000
Octapole moment (Debye-Ang**2):
   XXX =    .0000  YYY =     .0000  ZZZ =  -.2251   XYY =     .0000
   XXY =    .0000  XXZ =   -.0089   XZZ =    .0000   YZZ =     .0000
   YYZ =  -.5756   XYZ =     .0000
Hexadecapole moment (Debye-Ang**3):
  XXXX =  -3.2323 YYYY =   -6.4280 ZZZZ =  -4.8602 XXXY =     .0000
  XXXZ =    .0000 YYYX =     .0000 YYYZ =     .0000 ZZZX =     .0000
  ZZZY =    .0000 XXYY =   -1.7655 XXZZ =  -1.3841 YYZZ =  -1.6675
  XXYZ =    .0000 YYXZ =     .0000 ZZXY =     .0000
```

Note that the quantities are given in non-SI form; the unit of length is the Å, and the Debye (D) is a relic from the days of electrostatic (cgs) units, defined as the dipole moment corresponding to a pair of equal and opposite charges of magnitudes 10^{-10} esu, separated by 1Å. There are 2.998×10^9 esu per coulomb, and so $1\,D = 10^{-10}$ esu $\times 1 = 3.336 \times 10^{-30}$ C m. Also, $ea_0 = 8.4784 \times 10^{-30}$ C m, which is 2.5418 D.

13.2 IMPLICATION OF BRILLOUIN'S THEOREM

In each case, the electronic contribution to the properties involves a sum of one-electron operators $\Sigma x_k, \Sigma x_k^2$ etc. and Brillouin's theorem encourages us to believe that the calculation of such properties for a closed-shell molecule should be reliable at SCF level.

I mentioned Brillouin's theorem in our discussions of the excited states of π-electron systems; it refers to the SCF ground state of a molecule like H_2O, where we can describe the ground state as doubly occupied MOs, and informs us that singly excited states constructed from Hartree Fock orbitals do not interact with the electronic ground state

$$\int \Psi_0 H \psi(A \rightarrow X) \mathrm{d}\tau = 0$$

where the excitation is from an occupied SCF orbital ψ_A to the virtual orbital ψ_X. The argument is well-known; if we were to use CI to improve the ground state wavefunction by mixing in excited states, then according to perturbation theory the only states that would contribute would be doubly excited states and those with higher excitations. Analysis of the change that this would make in the predicted electrical moments shows why they are said to be 'correct to second order in perturbation theory for closed-shell molecules'. This conclusion is also valid for other one-electron properties.

A recent reference containing a good number of calculated values of electric multipole moments is that of Dykstra and Liu (1989).

It is usual to consider the following factors in discussions of one-electron properties.

- Relativistic effects
- The Born–Oppenheimer approximation
- Variation with bond distance
- Dependence upon basis set
- Correlation effects

Relativistic effects are generally thought to be negligible for 'light' nuclei. Spin orbit coupling is a relativistic effect that is well reported in tables of atomic energy levels, so my advice is this. If you intend to do ordinary SCF calculations (which are distinctly non-relativistic) on a complex involving Pt in the belief that your calculation will give a clue as to the causes and hence the cures for cancer, be aware that your calculated energy levels and one-electron properties may not turn out to be correct.

The Born–Oppenheimer approximation assumes that the nuclear and electronic motions are separable. Experimental estimates of the dipole moment error introduced by this approximation are readily available. For example, the small but non-zero permanent electric dipole moment of HD is 6×10^{-4} D (2×10^{-33} C m), and this is due entirely to the breakdown of the Born–Oppenheimer approximation (Trefler and Gush 1968). Otherwise, this effect is thought to be totally unimportant to chemical accuracy.

Basis set dependence is very important and Table 13.1 shows the effect of basis set on the calculated electric dipole of a sample molecule, pyridine.

Table 13.1 Electric dipole moment of pyridine

Level of theory	ϵ/E_h	$p_\text{e}/10^{-30}$C m
HF/STO-3G	−243.628920	6.502
HF/4-31G	−246.318523	9.015
HF/DZ	−246.588409	9.464
HF/DZV	−264.642356	9.080
Experimental		7.31±2%

There is a strong temptation when running *Ab Initio* packages to optimize the geometry of every molecule on every occasion, but it isn't always necessary. Geometry optimizations are very expensive calculations. I did the calculations above using the assumed geometry of a regular hexagon. As a general rule, at the *Ab Initio* level, minimal basis sets seem to underestimate electric dipole moments by a few per cent whilst extended basis sets usually overestimate them by a few per cent. Molecules containing fluorine seem to give problems in that the error is often large. Addition of polarization functions to extended basis sets usually reduces the calculated value.

Table 13.2 Electric dipole moment of CO

Level of theory	R_e/pm	$p_\text{e}/10^{-30}$ C m
HF/STO-4G OPT	114.6	+0.2165
HF/4-31G OPT	112.8	−2.0019
HF/6-311G OPT	112.4	−1.5896
HF/6-311G** OPT	110.5	−0.5111
MP2/6-311G** OPT	113.9	+0.1283
CID/6-311G** OPT	112.2	+0.2862
Experiment		+0.374±0.017

SCF calculations on molecules with small electric dipole moments need to be treated with caution.

The classic case is CO, and it won't do any harm to tell the familiar story once again. Burrus (1958) determined the magnitude of the vector from a Stark experiment as 0.112±0.005 D (0.374±0.017 10^{-30} C m). The polarity of C-O$^+$ was determined by

Rosenblum et al (1958) from a molecular beam electric resonance experiment. The C^-O^+ polarity goes entirely against simple chemical intuition.

Early *Ab Initio* HF calculations using minimal basis sets were spectacularly successful in that they gave the correct polarity, as Table 13.2 illustrates. Sadly, more extended basis sets give the wrong polarity at the HF level. Electron correlation is a must in this case. CO is either a spectacularly bad case at HF level, or an interesting opportunity for further study, depending on your viewpoint.

We should also consider the effect of molecular vibrations. The point is that the molecule is vibrating even at 0K, and many experimental measurements refer to an average over the vibrational state(s). Most theoretical calculations refer to the molecule in a hypothetical rotationless, vibrationless state, and obviously, the properties relating to the two are not the same.

The variation of a property P with bond distance R is usually discussed in terms of a power series $P(R) = P_0(R_e) + P_1(R - R_e) + P_2(R - R_e)^2 + \ldots$ and we can determine the coefficients $P_0, P_1 \ldots$ by calculating the property at different bond distances and then fitting the polynomial. The questions of how many points to take, whether to take the experimental R_e or the calculated one appropriate to that particular level of theory, and what degree of polynomial is appropriate all have to be addressed.

For illustration, I have calculated the dipole moment for CO at the HF/6–311G** level of theory for the range of bond distances, Table 13.3.

Table 13.3   Dipole moment vs. bond distance, HF/6-311G**, for CO

R/pm	$p_e/10^{-30}$ C m
95	1.8308
100	1.1222
105	0.3646
110.47 $= R_e$(calc)	−0.5111
115	−1.2767
120	−2.1137
125	−2.9787

At this level of theory, the calculated R_e is 110.47 pm, and fitting the calculated values to a quadratic in the bond extension by least squares gives the following equation

$$p_e/10^{-30} \text{ C m} = -0.5138 - 0.1616(R - R_e)/\text{pm} - 0.6009 10^{-4}(R - R_e)^2/\text{pm}^2$$

Incidentally, the cause of the problem concerning calculations of the absolute value of the dipole moment can be seen from these values. The dipole moment changes sign around R_e, and this explains why it is necessary to work so hard to get agreement with experiment for such a simple molecule. If we differentiate the expression with respect to R, we get the dipole derivative. Dipole derivatives give the intensities of infrared vibrational transitions, and it seems that the behaviour of p_e with R, given above, is correct.

You might like to read Green's review entitled 'Sources of error and expected accuracy in *Ab Initio* one-electron properties: the molecular dipole moment' (Green 1974).

Brillouin's theorem does not hold for open-shell electronic states, and the calculation of electric dipole moments is much less satisfactory in these cases. Some treatment of electron correlation is usually needed in order to obtain even modest agreement withal experiment results. For illustration, I have given a sample of calculations on CN in its lowest energy $X^2\Sigma^+$ state in Table 13.4.

Table 13.4   Electric dipole moment of CN in its $X^{2\ +}$ state

Level of theory	R_e/pm	$p_e/10^{-30}$ C m
HF/STO-4G OPT	124.4	−5.995
HF/4-31G OPT	118.6	−7.4079
HF/6-311G OPT	115.4	−7.3088
HF/6-311G** OPT	115.4	−7.3088
MP2/6-311G** OPT	113.0	−7.0609
CID/6-311G** OPT	113.5	−7.0006
CISD/6-311G** OPT	114.9	−6.4446
Experiment		(−)12.3±0.7

Table 13.5   Quadrupole moment of pyridine using four different basis sets. The coordinate origin is the centre of nuclear charge

Theory level	$\Theta/10^{-40}$ C m^2		
	Θ_{aa}	Θ_{bb}	Θ_{cc}
HF/STO-3G	−6.6219	20.0207	−13.3988
HF/4-31G	−19.4722	33.4409	−13.9687
HF/DZ	−21.3170	35.4157	−14.0987
HF/TZV	−21.1750	34.6943	−13.5189
Experiment	−21±5	33±4	−12±3

13.3   QUADRUPOLE MOMENTS

Molecular electric quadrupole moments are more elusive animals, and they are not particularly easy to measure experimentally. Prior to 1970 the only direct routes to these quantities were from the Kerr and the Cotton–Mouton effects. They can now be obtained from molecular microwave Zeeman spectroscopy, to a fair accuracy. It is true to say that calculation offers a faster route to this property than experiment, and perhaps a more reliable route.

Table 13.5 shows the sensitivity of quadrupole moment to the choice of basis set for pyridine. Apart from the minimal basis set calculation, all calculated values lie within the experimental error bars. The experimental quantities are not to be regarded as 'correct' any more than the *Ab Initio* values are 'correct'. Several experimental values in the literature refer to measurements corrected neither for the zero-point vibrations nor for centrifugal effects.

Incidentally, the output from GAUSSIAN92 shows second moments, despite being labelled as the quadrupole moments.

13.4 HIGHER ELECTRIC MOMENTS

Electric moments beyond the second are rarely encountered in chemistry. They do appear in the field of collision-induced spectroscopy, and there is no particular reason to assume that the experimental values are any more reliable than the calculated ones.

13.5 ELECTRIC FIELD GRADIENTS

I mentioned electron spin in the introductory chapters; electron spin is an intrinsic angular momentum that an electron possesses simply because it is an electron. It is part of the nature of things. The laws of quantum mechanics tell us three things about the electron spin angular momentum vector $\underline{s}$:

- the magnitude of $\underline{s}$ is $\sqrt{s(s+1)}h/2\pi$ where $s = \frac{1}{2}$;
- only the magnitude of $\underline{s}$ and that of one component (s_z) can be measured simultaneously;
- the z component of the vector has magnitude $s_s\, h/2\pi$.

Many **nuclei** have a corresponding internal angular momentum which we refer to as nuclear spin and we use the symbol $\underline{I}$ to represent the vector. The nuclear spin quantum number I is characteristic of the given nucleus and can have different values for different isotopic species. Many nuclei with nuclear spin quantum number $I \geq 1$ also posess a nuclear quadrupole moment. Nuclear physicists give it the symbol $\underline{\mathbf{Q}}_N$ (or sometimes just $\underline{\mathbf{Q}}$), and it is usually defined in the same way as the electric quadrupole moment discussed above

$$\underline{Q}_N = \frac{1}{e}\begin{pmatrix} \int \rho_N(\underline{r})(3x^2 - r^2)d\tau & 3\int \rho_N(\underline{r})xyd\tau & 3\int \rho_N(\underline{r})xzd\tau \\ 3\int \rho_N(\underline{r})yxd\tau & \int \rho_N(\underline{r})(3y^2 - r^2)d\tau & 3\int \rho_N\underline{r})yzd\tau \\ 3\int \rho_N(\underline{r})zxd\tau & 3\int \rho_N(\underline{r})zyd\tau & \int \rho_N(\underline{r})(3z^2 - r^2)d\tau \end{pmatrix}$$

apart from the $\frac{1}{2}$ and the division by e. Be aware though that some authors work with the definition

$$\frac{1}{e}\begin{pmatrix} \int \rho_N(\underline{r})x^2d\tau & \int \rho_N(\underline{r})xyd\tau & \int \rho_N(\underline{r})xzd\tau \\ \int \rho_N(\underline{r})yxd\tau & \int \rho_N(\underline{r})y^2d\tau & \int \rho_N\underline{r})yzd\tau \\ \int \rho_N(\underline{r})zxd\tau & \int \rho_N(\underline{r})zyd\tau & \int \rho_N(\underline{r})z^2d\tau \end{pmatrix}$$

If in doubt, check the sum of the diagonal elements. For the first definition, the sum is zero. For the second definition, it isn't!

In principle, if we know the nuclear charge density $\rho_N(\mathbf{r})$ then we could calculate the nuclear quadrupole moment in a similar way to the electric quadrupole, but in practice we have to determine nuclear quadrupole moments by experiment. The integration is now over the coordinates of the nuclear wavefunction. The charged particles in the nucleus are, of course, protons and they can be thought of as rotating very rapidly about the local z axis of the nuclear spin vector. If an average is made over a time long enough for the nuclear particles to rotate but so short that the electrons have not appreciably changed position, the nuclear charge distribution may be considered cylindrical. Using a principal axis system which coincides with the nuclear spin, all non-diagonal components of the nuclear quadrupole moment $\mathbf{Q}_N$ are zero and $Q_{N,xx} = Q_{N,yy} = -\frac{1}{2}Q_{N,zz}$. The entire quadrupole tensor is expressed in terms of 'the' value Q (or Q_N) $= Q_{N,zz}$.

In a molecule, a given nucleus will generally experience an electric field gradient due to the surrounding electrons, and simple arguments suggest that only p and d atomic orbitals can contribute to these field gradients.

The energy of interaction U between the nuclear quadrupole and the electric field gradient is given by

$$U = -\frac{e}{6}\sum\sum \underline{\mathbf{Q}}_{N,ij}\underline{\mathbf{E}}'_{ij} - \cdots$$

It is usual to denote the field gradients at the nuclear position by the matrix $\underline{\mathbf{q}}_N$, which is again a 3×3 quantity.

$$\underline{\mathbf{q}}_N = \begin{pmatrix} q_{xx} & q_{xy} & q_{xz} \\ q_{yx} & q_{yy} & q_{yz} \\ q_{zx} & q_{zy} & q_{zz} \end{pmatrix}$$

Yet another pit for the unwary, this property is **usually** defined in such a way that the sum of the diagonal elements is zero, but it doesn't have to be.

In what follows, I will assume a traceless electric field gradient tensor. In principal axes, the interaction is characterized by $Q_{N,zz}$ and two of q_{xx}, q_{yy} and q_{zz}. The largest of q_{xx}, q_{yy} and q_{zz} is given the symbol q, and we refer to eqQ/h as the **quadrupole coupling constant**. The quantity $\nu =$ (difference of smaller q's divided by the largest q) is called the asymmetry parameter.

We are not there yet. In order to find the allowed energy levels for a nuclear quadrupole in an external field gradient (which is produced by the molecular electrons), we need to

- write down the classical energy expression;
- deduce the correct Hamiltonian operator;
- solve the eigenvalue problem.

The interaction energy depends on Q_N, q and ν, and also on the relative orientations of the principal axes of the two matrices. To cut a long story short, the allowed energy levels turn out to depend on eQ_Nq. This is obviously an energy term and if we divide by h we arrive at the **quadrupole coupling constant**.

13.5.1 Experimental determination of eQNq/h

Quadrupole coupling constants for molecules are usually determined from the hyperfine structure of pure rotational spectra or from electric beam and magnetic beam resonance spectroscopies. Nuclear magnetic resonance, electron spin resonance and Mossbäuer spectroscopies are also routes to the property. A large amount of experimental data exists for ^{14}N and halogen-substituted molecules. There is less available data for deuterium because the nuclear quadrupole is small.

The quadrupole coupling constant (QCC) contains essentially two unknowns, the electric field gradient at a nuclear position and the nuclear quadrupole moment. For a simple atom or an anion, it is possible occasionally to make a realistic estimate of the field gradient, and so the QCC has sometimes been used to give a measure of the nuclear quadrupole moment. In a molecule, the nuclei are embedded in the electron cloud. The electric field gradient gives a measure of the departure from spherical symmetry of the electric charge distribution at the nucleus in question. It depends on the environment of the nucleus and is thus intimately connected with the bonding.

The chemical significance of nuclear quadrupole moments has been summarized by Townes and Dailey (1949), and reviewed by Orville-Thomas (1957). Dailey, in particular, points out that the quadrupolar nuclei can act as chemical probes that give information about the electron density at points inside the molecule. You have probably come across the Townes and Dailey interpretation, which runs as follows. Since filled shells and s orbitals have spherical symmetry, and since d and f orbitals do not penetrate near the nucleus, the quadrupole coupling should be largely due to any p electrons present in the valence shells. An interesting application is given in Orville-Thomas' paper, which has both an intriguing title and abstract;

> ### Clash between Experimental Parameters and Simple Theoretical Concepts
> #### W. J. Orville-Thomas
> *Journal of Chemical Physics*, **43** (1965) S244–S247
>
> *Facts are better than Dreams*
> – Winston Churchill

The author concerns himself with the comparison between theoretical and experimental studies of simple molecules containing the C≡N linkage.

A large number of spectroscopic studies have been made for XCN molecules (with a range of substituent X's), and in the majority of cases many isotopic species have been studied so that the experimental data is extremely accurate. An interesting feature of the experimental geometries is that the CN distances are equal to within experimental error, and one might expect that the ^{14}N quadrupole coupling constants would be very similar. The FCN QCC turns out to be very different from the others, which shows that bonds of equal lengths do not have the same electron distribution. The various principles are shown in Table 13.6.

Table 13.6   Experimental results

Molecule	$R(C≡N)$/Å	^{14}N QCC/MHz
HCN	1.155	−4.58
FCN	1.159	−2.67
ClCN	1.159	−3.63

The ^{14}N QCCs are obviously very different and, if anything, the ^{19}F value is 'anomalous'.

In order to model this, I ran HF/6–31G optimizations on the three molecules, and then calculated the electric field gradients at the nuclei Table 13.7.

Table 13.7   Theoretical HF/6-31G results

Molecule	$R(C≡N)$/Å	^{14}N QCC/MHz
HCN	1.144	−3.86
FCN	1.141	−2.20
ClCN	1.145	−2.82

I took the ^{14}N quadrupole moment as 0.016×10^{-28} m^2 and used the conversion factor that 1 au of electric field gradient $= 9.7174 \times 10^{21}$ V m^{-2} in order to convert the output from GAUSSIAN92. The agreement with experimental results leaves a little to be desired, but that is another story. The trend is well reproduced by the calculations.

As a general rule, minimal basis sets seem to give a rather poor representation of q and it is wise to stick to extended sets. Asymmetry factors are very elusive animals. The same comments, obviously, apply for nuclei other than ^{14}N. There is, for example, a large amount of data on deuterium and halogen quadrupole coupling in the literature. Deuterium data is hard to obtain experimentally because the nuclear quadrupole is small and because hydrogen atoms in molecules are almost spherical. In order to measure a molecular quadrupole moment, we need a very accurate strong electric field gradient, and these have proved difficult to generate experimentally.

13.6 THE ELECTROSTATIC POTENTIAL

One of the fundamental objectives of chemistry is the prediction and rationalization of molecular reactivity. In principle, this involves the calculation of a potential energy surface in the first instance. A great deal of labour and expense is involved normally in constructing *Ab Initio* potential energy surfaces to chemical accuracy for even simple reactions, and it is probably fair to say, in summary, that such predictions have so far failed to materialize, for everyday reactions, at the *Ab Initio* level. Most of the traditional theories of chemical reactivity have been aimed at organic molecules and these theories usually attempt to extract, from the electronic properties of an isolated molecule, some useful information as to how the molecule will interact with other molecules.

We can distinguish between **static** theories, which in essence give a description of the electronic wavefunction and **dynamic** theories which aim to predict the response of a molecule to (e.g.) an approaching point charge. In recent years, the electrostatic potential has been used to give a simple static representation of the outstanding features of molecular reactivity. I showed you earlier how to calculate the electrostatic potential at points in space due to point charges and charge distributions. I also explained how to calculate one-electron properties. In the case of the electrostatic potential, we have

$$\varphi(\mathbf{r}) = \frac{e}{4\pi\epsilon_0}\left(\sum_{\alpha=A}^{NUC}\frac{Z_\alpha}{R_\alpha} - \int P_1(\mathbf{r})\left(\frac{1}{r}\right)d\tau\right)$$

where the first term in brackets is the nuclear contribution and the second term is the contribution from the electron density. The electrostatic potential operators weight the inner regions of the wavefunction, since they depend on $1/r$, whilst the moment operators weight the outer regions since they depend on r, r^2, etc.

Generally, one presents calculations of the electrostatic potential as contour diagrams and to make our discussion concrete I will give several simple examples. It is usual to present contour diagrams of $+1C \times V(\mathbf{r})$ i.e. the work done in bringing a unit positive charge to position $\mathbf{r}$. The maps are, therefore, maps of mutual potential energy, but authors generally refer to them as potential energy maps. The spirit of this kind of calculations is to give a rough description of molecular reactivity. Even the molecular mechanics packages such as DTMM give a facility for the visualization of electrostatic potentials. For example, Figure 13.3 shows cholesterol, from the DTMM database.

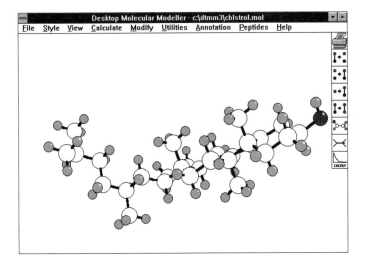

Figure 13.3   DTMM3 Windows screen showing cholesterol.

You should recognize the OH group in the top right-hand corner.

DTMM, like other molecular modelling packages, will calculate the electrostatic potential at points on the van der Waals' atomic surfaces, and represent this by shading. Here is cholesterol again, as processed by DTMM (Figure 13.4). Well, it looks better in colour than in monochrome!

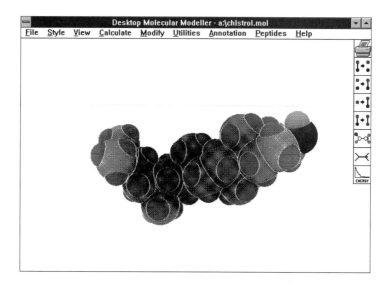

Figure 13.4   DTMM3 Windows screen showing the electrostatic potential on the van der Waals' surfaces

Other modelling packages such as MOBY allow the facility for calculating and visualizing electrostatic potential contours. The MOBY textbook example is *p*-aminobenzaldehyde, shown in Figure 13.5. What I did was to optimize the geometry using the internal AMBER forcefield, then run an AM1 energy calculation at this geometry. (In fact, MOBY will optimize a molecular geometry using AM1 and MNDO models, but I didn't take advantage of these facilities). I then calculated the electrostatic potential at the AMBER geometry using AM1, to give the figure. Again, it looks better in the original colour than in its present monochrome.

The general philosophy of this kind of work is that the electrostatic potential gives a rough-and-ready means of predicting reaction paths. It is not regarded as necessary to pay too much attention to the level of theory, and the procedure that I followed is perfectly adequate.

For a three-dimensional molecule, the visualization of these contours can present problems. I refer you to the review by Scrocco and Tomasi (1970) for more details.

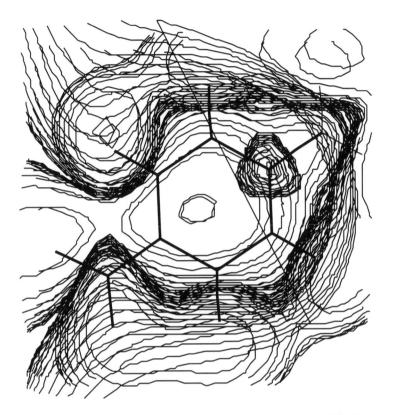

Figure 13.5 MOBY 1.5 representation of *p*-aminobenzaldehyde

14 Induced Properties

In the previous chapter, I showed you one possible way in which molecular properties could be classified. We studied the so-called primary properties, which I defined as those that could be obtained by direct calculation from the molecular wavefunction. It is now time to turn our attention to those properties that measure the response of a molecule to an external field. In Boys and Cook's language, these are the **induced properties**. As I mentioned earlier, this classification of properties is not unique, and the names can be misleading. For example, I will show you shortly that the electric dipole moment can be defined as the first derivative of a certain energy, and the mathematical technique of differentiation is often used to calculate induced properties.

14.1 ENERGY OF A CHARGE DISTRIBUTION IN A FIELD

Most molecular modelling is done within the Born–Oppenheimer approximation, where we regard the nuclei (point positive charges) as being fixed in space as far as the electron motions are concerned. The electrons are regarded as continuous charge densities, and the electrons generate the potential in which the nuclei vibrate. One way of studying electric charge distributions is to subject them to an external electric field. This usually changes their energy, and the change in energy can be studied by various experimental techniques. In spectroscopic experiments, the electric and magnetic fields involved are time dependent. This makes the treatment more complicated.

But to get you going in this topic, I will begin with a very simple physical model where the external electric field is constant with respect to time (it is a so-called **static** field).

Suppose once again that point charges $Q_A, Q_B, \ldots, Q_N$ are located at positions $\underline{r}_A, \underline{r}_B \ldots \underline{r}_N$ respectively. In the previous chapter, I showed you how to calculate the electric dipole moment for this charge distribution, and discussed the various definitions of the electric quadrupole moment. Suppose now we switch on an external field, defined at points $\underline{r}$ in space by the scalar potential $\varphi(\underline{r})$ (or the electric field $\underline{E}(\underline{r})$, since these two are related by $\mathbf{E} = -\mathrm{grad}\,\varphi$). The field need not be constant at the spatial positions, but it is taken to be time independent, for the moment.

In this context, I am going to follow normal convention and use the symbol W for energy rather than ϵ, and $\mathbf{E}$ for an electric field (which is a vector quantity). The energy of interaction W of the particles with the field is given exactly by

$$W = \sum_{i=A}^{N} Q_i \varphi(\mathbf{r}_i)$$

and the idea is to manipulate this equation in order to separate the properties of the charge distribution (such as the dipole moment) from the properties of the potential. What we do is to expand the potential calculated at each of the points r_i about the coordinate origin as a Taylor series to give

$$\varphi(\mathbf{r}_i) = \varphi_0 + \mathbf{r}_i.\mathrm{grad}\,\varphi + \text{higher terms}$$

where φ and its derivatives are all evaluated at the coordinate origin. The electrostatic field $\mathbf{E}$ and the potential are related by definition, $\mathbf{E} = -\mathrm{grad}\,\varphi$. This gives us expressions for the field gradient, and collecting together the results yields

$$W = Q\varphi_0 - \underline{\mathbf{p}}_e.\mathbf{E} - \frac{1}{3}\sum_{i=x}^{z}\sum_{j=x}^{z}\Theta_{ij}E'_{ij} - \dots$$

You should recognize the multipole moments in this expression.

So, in addition to being useful for the calculation of an electrostatic potential at points around a charge distribution, the multipole moments also determine the interaction energy of the charge distribution with an external field. In an external **uniform** field, the interaction energy of a neutral system of fixed point charges is given exactly by

$$W = -\underline{\mathbf{p}}_e.\underline{\mathbf{E}}$$

Consider now the simple case where the electric dipole moment and the external field both point along the same axis. They are related by

$$p_e = -\frac{\mathrm{d}W}{\mathrm{d}E}$$

This shows how the electric dipole moment can be defined as the first derivative of the interaction energy when a charge distribution is placed in an external field. For that reason, several authors talk about electric moments as derivative properties rather than primary ones. They are given by the derivative of the energy with respect to the dipole moment.

A special feature of the quadrupole moment term is that the components of the field gradient are not all independent, for the following reason. The electrostatic potential φ satisfies Laplace's equation

$$\frac{\partial^2 \varphi}{\partial x^2} + \frac{\partial^2 \varphi}{\partial y^2} + \frac{\partial^2 \varphi}{\partial z^2} = 0$$

or $\quad E'_{xx} + E'_{yy} + E'_{zz} = 0$.

The interaction energy is therefore unchanged if any constant is added to the diagonal elements of the second moment tensor, and this explains the various definitions of the electric quadrupole moment. As I mentioned earlier, modern usage seems to dictate that we should work directly with the moments such as x^2.

14.2 INDUCED DIPOLES

Molecules do **not** consist of rigid arrays of point charges, and on application of an external electrostatic field the electrons and protons will rearrange themselves until the interaction energy is a minimum. In classical electrostatics, where we deal with macroscopic samples, this phenomenon is referred to as the **induced polarization**. The nuclei and electrons tend to move in opposite directions, since they are oppositely charged, and the molecular electric moments will therefore change.

Since we normally have to deal with electrostatic field perturbations that are small, we expand the electric multipole moments $\underline{p}_e \dots$ as a Taylor series in the external field $\mathbf{E}$ as follows

$$\underline{p}_e(\mathbf{E}) = \underline{p}_e(\mathbf{E} = \mathbf{O}) + \underline{\alpha}.\mathbf{E} + \dots$$

The electric dipole moment $\mathbf{P}_e(\mathbf{E} = \mathbf{O})$ is of course the **permanent electric dipole moment**. The property $\underline{\alpha}$ is a **dipole polarizability**, and can be written as a 3×3 symmetric matrix

$$\underline{\alpha} = \begin{pmatrix} \alpha_{xx} & \alpha_{xy} & \alpha_{xz} \\ \alpha_{yx} & \alpha_{yy} & \alpha_{yz} \\ \alpha_{zx} & \alpha_{zy} & \alpha_{zz} \end{pmatrix}$$

Both $\underline{p}_e(\mathbf{E} = \mathbf{O})$ and the applied field $\mathbf{E}$ are vector quantities, and the direction of the induced dipole may not be the same as the direction of the applied field. Hence, we need a more general property than a vector in order to describe the polarizability.

The polarizability, the field gradient and the quadrupole moment are all examples of **tensor properties**. A detailed treatment of these tensor properties is outside the scope of this text, but you should be aware of the existence of such entities. For the special case of an axially symmetric molecule, the electric dipole moment lies along the axis of highest symmetry which we usually call the z axis. The induced dipole due to a field along the molecular axis is usually written

$$\underline{p}_{e,z} = \underline{p}_e + \alpha_{zz}\mathbf{E}_z$$

whilst for a field perpendicular to the axis we have

$$\underline{p}_{e,x} = \alpha_{xx}\mathbf{E}_x$$

The effect of the induced polarization on the energy of the charge distribution in a uniform electrostatic field can be deduced as follows. We have

$$\underline{\mathbf{p}}_e = -\frac{\mathrm{d}W}{\mathrm{d}E}$$

and

$$\underline{\mathbf{p}}_e(\mathbf{E}) = \underline{\mathbf{p}}_e(\mathbf{E} = \mathbf{O}) + \underline{\alpha}.\mathbf{E} + \ldots$$

Combining these two and integrating gives

$$W = W_0 - \underline{\mathbf{p}}_{e,0}.\mathbf{E} - \frac{1}{2}\sum_{i=x}^{z}\sum_{i=x}^{z} E_i\,\alpha_{ij}\,E_j - \ldots$$

where W_0 is the energy in the absence of a field.

The cartesian components of the electric dipole and of the electric quadrupole can therefore be written

$$(\underline{\mathbf{p}}_{e,0})_i = -\left(\frac{\mathrm{d}W}{\mathrm{d}E_i}\right)_{\mathbf{E}=\mathbf{O}}$$

and

$$\alpha_{ij} = -\left(\frac{\mathrm{d}^2 W}{\mathrm{d}E_i\mathrm{d}E_j}\right)_{\mathbf{E}=\mathbf{O}}$$

which show how we can relate the permanent dipole moment to the first derivatives of the energy at zero field, and the dipole polarizability to the second derivatives of the energy, again at zero field.

The latter equation demonstrates that the dipole polarizability tensor is symmetric $\alpha_{ij} = \alpha_{ji}$ and so there are no more than six independent components of the dipole polarizability matrix.

As with all such tensor properties, we can find a set of coordinate axes labelled a, b and c such that the matrix has a diagonal form with respect to these axes

$$\underline{\alpha} = \begin{pmatrix} \alpha_{aa} & 0 & 0 \\ 0 & \alpha_{bb} & 0 \\ 0 & 0 & \alpha_{cc} \end{pmatrix}$$

where α_{aa}, α_{bb} and α_{cc} are the **principal values of polarizability**. For a molecule with symmetry axes, the principal axes correspond to the symmetry axes.

For a linear molecule, it should be obvious from symmetry that the effect of an applied electric field perpendicular to the molecular axis will be the same irrespective of any angular dependence, and the polarizability tensor takes an even simpler form

$$\underline{\alpha} = \begin{pmatrix} \alpha_| & 0 & 0 \\ 0 & \alpha_\perp & 0 \\ 0 & 0 & \alpha_\perp \end{pmatrix}$$

Experimental measurements usually focus on the **mean value**

$$\bar{\alpha} = \frac{1}{3}(\alpha_| + 2\alpha_\perp)$$

and the **anisotropy**

$$\beta = (\alpha_| - \alpha_\perp)$$

which is usually a positive quantity.

Experimental determinations are far from straightforward, especially if the molecule has little or no symmetry. The mean value can be obtained from the refractive index of a gas, whilst Kerr experiments give some idea of the anisotropy.

14.2.1 Multipole polarizabilities

Just as the dipole moment changes in the presence of an external field, so the other electric moments also change and we can discuss the quadrupole polarizability, if we so wish. The theory is the same, and we would write the electric quadrupole moment as a Taylor expansion in terms of the applied field.

14.2.2 A classical model of dipole polarizability

An insight into the physical mechanism of polarizability can be provided by studying the response of a very simple 'atom' to an applied electric field. Our model of the atom has a point nuclear charge $+Q$ surrounded by a uniform spherical distribution of negative charge. This model atom is taken to have radius 'a', and the negative charge $-Q$ fills the surrounding sphere with uniform density. We now apply an external electric field $\underline{E}$ pointing in the direction shown, which I have called the z axis. The nucleus will be displaced a certain distance d in the $+z$ direction, as shown in Figure 14.1.

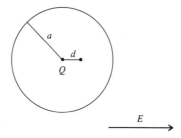

Figure 14.1   Construct needed to discuss the polarizability of a charge distribution

At this point the force on the nucleus has magnitude QE due to the applied electric field **E**, and the force due to the electron density, which can be shown (e.g. by Gauss' theorem) to be $Q^2d/4\pi\epsilon_0 a^3$. At equilibrium, these two forces must be equal and so the displacement d satisfies the equation.

$$\frac{Q^2d}{4\pi\epsilon_0 a^3} = QE$$

The induced dipole moment is Qd and since the polarizabilty is Qd/E

$$\alpha = 4\pi\epsilon_0 a^3$$

Apart from the factor of $4\pi\epsilon_0$, the polarizability of an atom is determined in this model by the atomic volume. If we had been working in cgs, we would have found $\alpha = a^3$ without the $4\pi\epsilon_0$. For this reason, people often speak of the atomic polarizability as a volume property, and persist in using cgs units for its measure.

The polarizability of the inert gas atoms does indeed increase with their size, as can be seen from Table 14.1. and you should keep this relationship at the back of your mind when discussing polarizability data. The SI units of polarizability are $F\,m^2$. Alternatively, since $F\,m^{-1} = C^2\,m^{-2}\,N^{-1}$ we can quote polarizabilities as $C^2\,N^{-1}\,m$ or finally $C^2\,m^2\,J^{-1}$, which is the normal form. In the atomic system of units, it is usual to quote polarizabilities in multiples of $e^2 a_0^2 E_h^{-1}$.

Table 14.1   Atomic polarizabilities

Atom	$\alpha/10^{-40}\,F\,m^2$
He	0.23
Ne	0.44
Ar	1.83

Atomic and molecular polarizabilities are usually determined experimentally by measuring the response of a system to electromagnetic radiation. It turns out that the measured values depend on the frequency of the radiation used; they are said to be **frequency dependent** properties.

Roughly speaking, it is observed experimentally that the dipole polarizability increases as the frequency of radiation increases. Eventually though, the energy of each photon is sufficient to cause an electronic transition, and in this case we see a large change in the polarizability. In order to account for the photon absorption, the model chosen has to take account of the quantum mechanical nature of matter. Our simple model discussed above does **not** take this into account.

Molecules therefore respond to frequency dependent fields in a quite different way than to static fields, and some care has to be exercised when comparing the results of theory and experiments. Ignore then the possibility that the photon might be absorbed, due to a favourable energy level separation.

To give you a simple classical model for frequency dependent polarizabilities, let me go back to Figure 14.1. In the static field case, the restoring force on the displaced nucleus turns out to be proportional to the displacement d,

$$\text{Restoring force} = \frac{Q^2}{4\pi\epsilon_0 a^3}d$$

and so this corresponds to a simple harmonic oscillator with force constant

$$k = \frac{Q^2}{4\pi\epsilon_0 a^3}$$

Since nuclear masses are much greater that electron masses we can treat the nucleus as if it were essentially fixed in space. Taking the mass of the charge cloud as m, then $k = m\omega_0^2$ where ω_0 is the radian frequency of the oscillator. Comparing equations, we find that

$$\alpha = \frac{Q^2}{m\omega_0^2}$$

where α is the polarizability of our model atom in the presence of a constant electric field.

A simple time-dependent electric field can be written as $\mathbf{E}(\mathbf{r}, t) = \mathbf{E}_0 \exp[j\omega t]$ where $E(r)$ depends on position but not on time, is the radian frequency of the field and j is the square root of -1 ($j^2 = -1$).

The polarizability of a similar harmonic oscillator in the presence of such a field characterized by ω turns out to be

$$\alpha(\omega) = \frac{Q^2}{m(\omega_0^2 - \omega^2)}$$

This simple classical model shows that the polarizability increases as ω increases, in agreement with experimental results. (As I mentioned above, the model cannot cope with electronic transitions when the frequency of the applied field is such that the molecular energy difference is equal to the photon energy).

14.2.3 Calculations of static dipole polarizabilities

The first step in any quantum mechanical model is the construction of an appropriate Hamiltonian. If we look back at the classical expression for the energy of a charge distribution in the presence of an external electrostatic field, then for a **weak** field (with no E^2 terms)

$$W = W_0 - \underline{\mathbf{p}}_{e,0}.\underline{\mathbf{E}}$$

and it is intuitively obvious how we should modify a corresponding field-free Hamiltonian. We just add on the extra terms corresponding to the field giving the operator

$$H = H_0 - \underline{\mathbf{p}}_e.\underline{\mathbf{E}}$$

Dipoles and electric fields are vector quantities, and we need to cater for this. What we often do is to add in turn an electric field along three mutually perpendicular axes and redo the standard calculation. There are several ways in which we can proceed, and a history lesson might be again in order.

Standard perturbation theory

I discussed the concepts of perturbation theory when dealing with electron correlation in Chapter 10. Suppose that we were in the lucky position of having been able to solve exactly a field-free case

$$H_0 \psi_{0,i} = W_{0,i} \Psi_{0,i}$$

where i labels the states, of which there could be an infinite number. (Remember that I am using W for energy in this chapter, to avoid confusion with the electric field E). More to the point, the Hamiltonian could be the Hartree Fock Hamiltonian, in which case we do know the exact soutions to this problem.

Perturbation theory is treated in all the standard texts on quantum mechanics, and gives us techniques for writing the solutions Ψ_i and ϵ_i of the related problem

$$(H_0 + \lambda\Delta)\Psi_i = W_i\Psi_i$$

in terms of the solutions for the unperturbed problem. Δ is called the perturbation and λ is a parameter included to keep track of the orders of magnitude. The dipole polarizability components can be deduced in this formalism as (for example)

$$\alpha_{xx} = 2\sum_{m=1}^{i} \frac{|\int \Psi_m p_x \Psi_0 d\tau|^2}{W_m - W_0}$$

where p_x is the x component of the dipole moment operator. This expression is not much help for practical calculation as it needs information about all the excited states in addition to the ground state.

Self consistent perturbation theory

In the context of Hartree Fock self consistent field theory with an LCAO expansion of basis functions $\chi_1, \chi_2, \ldots \chi_n$, you should be familiar by now with solving the HF equations which I wrote in Chapter 5 as

$$\underline{\mathbf{h}}^F \underline{\mathbf{a}} = W_A \mathbf{S}\underline{\mathbf{a}}$$

The idea of self consistent perturbation theory is to seek solutions of the perturbed HF equations

$$(\underline{\mathbf{h}}^F + \lambda\underline{\Delta})\underline{\mathbf{a}} = W_A \mathbf{S}\underline{\mathbf{a}}$$

in such a way that the orders of perturbation are kept separate. The dipole polarizability is calculated directly from the second-order energy. This technique was introduced by McWeeny.

The finite field SCF technique

This is similar to the self consistent perturbation approach, except that we just solve the modified HF equations

$$(\underline{\mathbf{h}}^F + \Delta)\underline{\mathbf{a}} = W_A \mathbf{S}\underline{\mathbf{a}}$$

for some arbitrary, small electric field without bothering to separate out all the orders of perturbation theory. Imagine then a very simple finite field calculation at the HF level of theory on CH_3F. We run a calculation on the molecule, then include, in order, an external electric field along each of the three cartesian axes. We then look either at the induced dipole or at the change in energy in order to deduce the dipole polarizability components.

In the LCAO approach with basis functions $\chi_1, \chi_2 \ldots \chi_n$ this means calculating an extra matrix of dipole integrals over the basis functions e.g. $\mathbf{X}$ when the applied field points in the x direction. We then add $-E\mathbf{X} = \underline{\delta}$ to the one-electron part of the Hartree Fock matrix. Mentally, we keep a record of the nuclear contribution to the dipole moment, but don't vary the nuclear geometry.

In practice, the calculation is not quite so simple because the higher-order terms in the induced dipole and energy expressions are not usually negligible. So, normally, we take a number of applied fields along each axis, typically multiples of $\pm 5 \times 10^{-4}$ au, and use standard techniques of numerical analysis in order to extract the required data. It is usual to use the energy expression rather than the induced dipole expression for this purpose.

The problem with such finite field calculations is that the evaluation of the dipole polarizability requires several energy evaluations, at least three in the simple case ($E = 0$ and $E = \pm 5 \times 10^{-4}$ au). Such calculations are costly and are rarely used today.

Energy derivatives

An alternative is to evaluate the energy derivatives directly as analytical or numerical expressions, which we then solve for the properties of interest. A great deal of effort has been devoted to finding analytical expressions at varying levels of theory over the last couple of decades.

14.2.4 A GAUSSIAN92 run

GAUSSIAN92 includes code to calculate force constants, polarizabilities and dipole derivatives analytically for HF, MP2 and other levels of theory, and numerically for the remaining options. For the record, here is the (edited) output file from a GAUSSIAN92 run on SO_2 at the HF/4–31G level of theory. I first optimized the geometry in a separate run, assuming C_{2v} symmetry, and then included the control word 'POLAR' to give the following.

SULFUR DIOXIDE OPTIMIZED GEOMETRY

Symbolic Z-matrix:
  Charge = 0 Multiplicity = 1
S
O1   S RSO
O2   S RSO O1 ANG
  Variables:
RSO   1.5284
ANG   114.3673

Z-Matrix (ANGSTROMS AND DEGREES)

CD	Cent	Atom	N1	Length/X	N2	Alpha/Y	N3	Beta/Z	J
1	1	S							
2	2	O	1	1.528400(1)					
3	3	O	1	1.528400(2)	2	114.367(3)			

Z-Matrix orientation:

Center Number	Atomic Number	Coordinates (Angstroms)		
		X	Y	Z
1	16	.000000	.000000	.000000
2	8	.000000	.000000	1.528400
3	8	1.392249	.000000	−.630594

STOICHIOMETRY O2S
FRAMEWORK GROUP   C2V[C2(S),SGV(O2)]
DEG. OF FREEDOM 2
FULL POINT GROUP   C2V NOP 4
LARGEST ABELIAN SUBGROUP   C2V NOP 4
LARGEST CONCISE ABELIAN SUBGROUP   C2 NOP 2

Standard orientation:

Center Number	Atomic Number	Coordinates (Angstroms)		
		X	Y	Z
1	16	.000000	.000000	.414157
2	8	.000000	1.284486	−.414157
3	8	.000000	−1.284486	−.414157

Rotational constants (GHZ): 46.0645185 9.5751767 7.9273602
Isotopes: S-32,O-16,O-16
Standard basis: 4-31G (S, S = P, 6D, 7F)
There are 14 symmetry adapted basis functions of A1 symmetry.
There are   2 symmetry adapted basis functions of A2 symmetry.
There are   5 symmetry adapted basis functions of B1 symmetry.
There are 10 symmetry adapted basis functions of B2 symmetry.

Closed shell SCF:
Requested convergence on RMS density matrix = 1.00D−08 within 64 cycles.
Requested convergence on MAX density matrix = 1.00D−06.
Two-electron integral symmetry used by symmetrizing Fock matrices.

SCF DONE: E(RHF) = −546.368405498   A.U. AFTER 13 CYCLES
      CONVG = .6683D-08   −V/T = 1.9999
      S**2 = .0000

Minotr: Closed-shell wavefunction.
      Direct CPHF calculation.
      Solving linear equations simultaneously.
      Using symmetry in CPHF.

      Differentiating once with respect to electric field.
        with respect to dipole field.

Symmetry not used in FoFDir.
      There are 3 degrees of freedom in the 1st order CPHF.
 3 vectors were produced by pass 0.
AX will form 3 AO Fock derivatives at one time.
 3 vectors were produced by pass 1.
 3 vectors were produced by pass 2.
 3 vectors were produced by pass 3.
 3 vectors were produced by pass 4.
 3 vectors were produced by pass 5.
 3 vectors were produced by pass 6.
 3 vectors were produced by pass 7.
 2 vectors were produced by pass 8.
 2 vectors were produced by pass 9.
Inv2: IOpt = 1 Iter = 1 AM = 1.75D-15 Conv = 1.00D-12.
Inverted reduced A of dimension 28 with in-core refinement.
Copying SCF densities to generalized density rwf, ISCF = 0
IROHF = 0.

****

Population analysis using the SCF density.

****

ORBITAL SYMMETRIES.
   OCCUPIED (A1) (A1) (B2) (A1) (B2) (B1) (A1) (A1) (B2) (A1)
      (A1) (B2) (B1) (B2) (A1) (A2)
   VIRTUAL (B1) (B2) (A1) (A1) (B1) (A1) (B2) (A1) (A2) (B1)
      (B2) (A1) (B2) (B2) (A1)
THE ELECTRONIC STATE IS 1-A1.
Alpha occ. eigenvalues −   −92.05384   −20.61240   −20.61239   −9.19753   −6.88315
Alpha occ. eigenvalues −   −6.87903   −6.87861   −1.49281   −1.38311   −.90835
Alpha occ. eigenvalues −   −.67946   −.66464   −.64798   −.52987   −.49303
Alpha occ. eigenvalues −   −.47687
Alpha virt. eigenvalues −   −.05073   .19494   .19915   .70655   .78017
Alpha virt. eigenvalues −   .79039   .80436   1.18597   1.22557   1.28632
Alpha virt. eigenvalues −   1.29948   1.30876   1.37259   1.73072   1.77741
      Condensed to atoms (all electrons):

		1	2	3
1	S	14.421564	.058907	.058907
2	O	.058907	8.724939	−.053534
3	O	.058907	−.053534	8.724939

Total atomic charges:
$$1$$
1 S   1.460622
2 O $-.730311$
3 O $-.730311$
Sum of Mulliken charges = .00000

Dipole moment (Debye):
  X = .0000   Y = .0000   Z = 3.2855   Tot = 3.2855

Exact polarizability: 8.699   .000   36.385   .000   .000   12.853

The polarizability tensor is given in the order $\alpha_{xx}, \alpha_{xy}, \alpha_{yy} \ldots \alpha_{zz}$ and the output is atomic units. An atomic unit of dipole polarizability is 1.6488×10^{-41} $C^2 m^2 J^{-1}$, so that the mean value

$$\bar{\alpha} = \frac{1}{3}(\alpha_{xx} + \alpha_{yy} + \alpha_{zz})$$

is 3.1842×10^{-40} $C^2 m^2 J^{-1}$. The experimental value quoted by Bogaard and Orr is 4.76×10^{-40} $C^2 m^2 J^{-1}$ as determined from measurements of the relative permittivities of bulk samples, or $4.326\ 10^{-40}$ $C^2 m^2 J^{-1}$ as determined from measurements made with an optical electric field corresponding to radiation of wavelength 632.8 pm. We should compare our calculated mean value with the value corresponding to zero frequency (the first experimental one quoted). Having said that, it is pretty clear that our theoretical value is in poor agreement with experiment.

For linear molecules, it is also possible to determine the anisotropy β where

$$\beta = (\alpha_{\parallel} - \alpha_{\perp})$$

but in the case of a non-linear molecule like SO_2, the experimental quantity of interest is a dimensionless quantity defined as

$$\kappa = \frac{\left(\sum_{i=x}^{z} \alpha_i^2\right) - 3\bar{\alpha}^2}{6\bar{\alpha}^2}$$

For a molecule with a three-fold or higher axis,

$$\kappa = \frac{\alpha_{\parallel} - \alpha_{\perp}}{3\bar{\alpha}}$$

The second thing to be aware of is the sensitivity of the calculated quantity to the level of theory, and I can best illustrate this for you in Table 14.2, which refers to calculations on SO_2. For the HF/STO–4G, the HF/4–31G and the HF/6–311G* calculations, I optimized the geometries before calculating the polarizabilities. For the remaining calculations, I left the geometry fixed at the HF/6–311G* level of theory values. I chose a simple molecule containing a second-row atom because there has been a great deal of controversy regarding the contribution, or otherwise, to the contribution made by d orbitals when predicting molecular structures. Geometries predicted by the simpest three models are given in Table 14.2.

Table 14.2 SO_2 geometry

Level of theory	R_{SO}/pm	OSO/°
HF/STO–4G	155.1	105.9
HF/4–31G	152.8	114.4
HF/6–311G*	140.7	118.7

There is, obviously, a large basis set effect in the prediction of molecular geometry. Bear this in mind for the polarizability calculations. I therefore optimized the geometries at the HF/STO–4G, the HF/4–31G and the HF/6–311G* levels before calculating the polarizabilities (Table 14.3). What happens normally is that the polarizabilities get larger as the basis set becomes more sophisticated, other things being equal. For a planar molecule, the 'perpendicular' component of the polarizability is often the most sensitive to improvements in the basis set. In this case, the perpendicular component corresponds to the x axis.

Table 14.3 Dipole polarizability/10^{-40} $C^2m^2J^{-1}$ of SO_2

Level	α_{xx}	α_{yy}	α_{zz}	Mean
HF/STO–4G	0.643	4.263	1.210	2.039
HF/4–31G	1.434	5.999	2.119	3.184
HF/6–311G*	1.993	4.447	2.313	2.918
HF/6–311++G(2d)	2.622	4.789	3.021	3.477
CID	2.647	4.993	3.174	3.605

The reason why the mean value falls on expansion of the basis set from the STO/4–31G level to the 6–311G* level is because of the dramatic effect of polarization functions on the geometry prediction. The CID results demonstrate that electron correlation effects are not particularly important for polarizability calculations.

14.3 INTERACTION POLARIZABILITIES

The virial equation of state is sometimes written

$$\frac{pV}{nRT} = A + BV_m^{-1} + CV_m^{-2} + \dots$$

This virial equation of state applies to all substances under all conditions of p, V_m and T. For an ideal gas, $A = 1$ and all the higher virial coefficients are zero. For other materials, the virial coefficients give information about molecular interactions. The constants $A, B, C \dots$ are temperature dependent but they are independent of density.

By analogy, the virial expansion for a bulk molecular property X (such as the dielectric polarization) is written

$$X = A_X + \frac{B_X}{V_m^1} + \frac{C_X}{V_m^2} + \cdots$$

where once again the constants A_X, B_X are dependent on temperature but independent of density.

Theoretical expressions for the virial coefficients can be found, in principle, by the application of statistical mechanics.

For static electric properties, the bulk property of interest is the dielectric polarization P, whose magnitude can be found (for example) from the Clausius Mossotti relation

$$P = \frac{\epsilon_r - 1}{\epsilon_r + 2} V_m$$

The dielectric virial equation for the polarization P is

$$P = A_\epsilon + \frac{B_\epsilon}{V_m^1} + \frac{C_\epsilon}{V_m^2} + \cdots$$

The molecular interpretation of this bulk equation is that A_ϵ represents the contribution of unperturbed molecules to the dielectric polarization, B_ϵ gives the pairwise interactions and so on. The highest-order term that is ever considered is B_ϵ.

The first dielectric virial coefficient can be found from the Clausius Mossotti equation in its original form, as obtained by Debye;

$$A_\epsilon = \frac{L}{3\epsilon_0} (\alpha_0 + \frac{p_e^2}{3kT})$$

The statistical mechanical expression for B_ϵ has been given by Buckingham as

$$B_\epsilon = \frac{4\pi L^2}{3\epsilon_0} \int_0^\infty \left\{ \left[\frac{1}{2}\alpha_{12}(r) - \alpha_0 \right] + \frac{1}{3}kT \left[\frac{1}{2}p_{12}^2(r) - p_0^2 \right] \exp\left[-\frac{W_{12}(r)}{kT} \right] \right\} r^2 dr$$

where $\alpha_{12}(r)$ is the polarizability of a pair of atoms separated by scalar distance r, p_e the electric dipole moment and W the potential energy. The pair polarization is an important quantity, though it has received much less attention than the potential energy.

An important experimental method of studying molecular interactions in gases and liquids is the scattering of laser light. When polarized light is scattered by a fluid, both polarized and depolarized components are produced. The depolarized spectrum is several orders of magnitude less intense than the polarized spectrum and, therefore, more difficult to observe. For simple fluids, much information about molecular motions has been obtained from these spectral analyses. A number of theories of the origin of the depolarized spectrum have been reviewed by Gelbart (1974). One of the simplest theories is the isolated binary collision (IBC) model due to McTague and Birnbaum.

In the IBC model, all effects due to the interaction of three or more particles are ignored and the scattering is due only to diatomic collision processes. There are only two unique components of the pair polarizability tensor (the parallel and the perpendicular components, referred to the internuclear axis), and attention focuses on the anisotropy $\beta(r)$ and the **incremental mean pair polarizability**

$$\bar{\alpha}^{(2)}(r) = \frac{1}{3}[\alpha_{\parallel}(r) + 2\alpha_{\perp}(r)] - (\alpha_A + \alpha_B)$$

where α_A and α_B are the polarizabilities of atoms A and B. In the IBC model, the polarized spectrum depends on the thermal average of this quantity over the whole assembly of atoms. The thermal average of $\beta(r)$ is also responsible for the depolarized spectrum.

The first *ab initio* study was that of O'Brien et al (1973) on a pair of helium atoms. They obtained $\beta(r)$ for the range $r = 3.5a_0$ through $10.0a_0$. The experimentally determined value of B_ϵ is negative, which suggests that the incremental mean pair polarizability must be negative around the minimum in the potential curve.

Alkali halides are important commercial compounds because of their role in extractive metallurgy. A deal of effort has, therefore, gone into corresponding calculations on alkali halides such as LiCl, with a view to understanding the structure and properties of ionic melts. Care has to be exercised in choosing a basis set, but calculations at the HF level of theory are usually judged to be adequate. Figures 14.2 and 14.3 show the variation of the anisotropy and the incremental mean pair polarizability for LiCl as a function of distance.

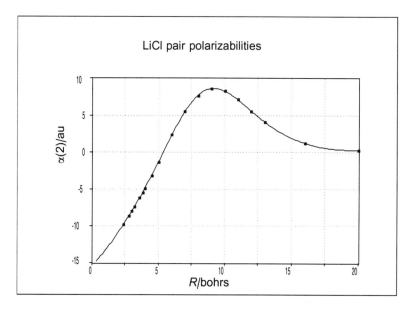

Figure 14.2   Pair polarizabilities for the LiCl ion pair

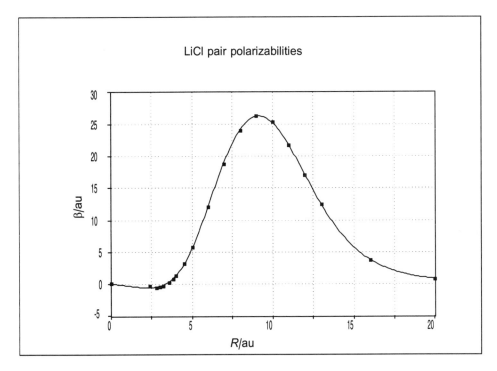

Figure 14.3   Pair anisotropy for LiCl

For most diatomic molecules, $\alpha | > \alpha \perp$. and at the limits $r = 0$ and $r = \infty, \beta = 0$. It proves useful to discuss the shapes of these curves in terms of a reduced distance $\sigma = r/r_{eq}$, and it turns out that the shapes of these curves are similar to the curves for the covalent molecule H_2.

14.4   THE HAMILTONIAN

I appealed to your good sense when I asked you to believe that the correct Hamiltonian for a molecule in an external electrostatic field $\underline{E}$ could be written

$$H = H_0 - \underline{p}_e.\underline{E}$$

In the case of electrostatic and magnetostatic fields, it is necessary to go back quite a way in order to develop the correct Hamiltonian. In electromagnetism, static fields are usually described by the electric field vector $\underline{E}$ and the magnetic induction $\underline{B}$. For our purposes, we have to be concerned with the potentials rather than the field, and these potentials are given as

$$\underline{E} = -\text{grad } \varphi$$

$$\underline{B} = \text{curl } \underline{A}$$

$\underline{A}$ is referred to as the **vector potential** and it is a fundamental construct in electromagnetism.

Imagine for the moment a free particle of charge Q and mass m. The Hamiltonian is

$$H_0 = \frac{p^2}{2m}$$

where p^2 is the square of the linear momentum operator $\mathbf{p}$. In the presence of external fields defined by the potentials φ and $\underline{\mathbf{A}}$, it turns out that the correct Hamiltonian is

$$H = \frac{|\mathbf{p} - Q\underline{\mathbf{A}}|^2}{2m} + Q\phi$$

The operator $\mathbf{p}$ is the momentum operator and the operator $\mathbf{p} - Q\underline{\mathbf{A}}$ is called a **generalized momentum** operator. If you know about the special theory of relativity, you might recognize the four quantities A_x, A_y, A_z and φ as the components of a **four-vector**. But that's not important right now.

In the Schrödinger picture, we put $\underline{\mathbf{p}} \rightarrow (-jh/2\pi)\nabla$ and after some manipulation we find

$$H = H_0 + \frac{jhQ}{2\pi m}\underline{\mathbf{A}}.\nabla + \frac{Q^2 A^2}{2m} + Q\phi$$

where $j^2 = -1$. The scalar potential $\underline{\mathbf{A}}$ is defined by $\text{curl}\,\underline{\mathbf{B}} = \underline{\mathbf{A}}$, and so it is undetermined to within a 'constant' of integration. In deriving this form for the Hamiltonian H, I have noted mentally that $\underline{\mathbf{A}}$ had to satisfy the condition that div $\underline{\mathbf{A}} = 0$. Some people talk about this as a **choice of gauge**. In the special case where $\underline{\mathbf{A}} = 0$ and the electrostatic field is uniform along the z axis, $\varphi = -Ez$ giving

$$H = H_0 - QEz$$

You should recognize Qz as a dipole moment operator.

In the case of a uniform magnetic induction $\underline{\mathbf{B}} = (0, 0, B)$, you will find that the vector field $\underline{\mathbf{A}} = -B(y, -x, 0)$ satisfies the requirement that $\underline{\mathbf{B}} = \text{curl}\,\underline{\mathbf{A}}$ and div $\underline{\mathbf{A}} = 0$. Substituting this expression for $\underline{\mathbf{A}}$ into the Hamiltonian yields

$$H = H_0 + \frac{jhQB}{4\pi m}\left(-y\frac{\partial}{\partial x} + x\frac{\partial}{\partial y}\right) + \frac{Q^2 B^2}{8m}(x^2 + y^2)$$

and you might recognize the first expression in brackets as it is obviously related to the z component of the angular momentum operator, L_z.

$$H = H_0 - \frac{QB}{2m}L_z + \frac{Q^2 B^2}{8m}(x^2 + y^2)$$

14.5 MAGNETIZABILITIES

The magnetic case is more complicated than the electrostatic case, because we have to deal with both of the terms and they enter the Hamiltonian with different powers of the magnetic induction B. So if we were to use perturbation theory to estimate the effects of these two terms, we would once again write the zero-field solution as

$$H_0 \Psi_{0,i} = W_{0,i} \Psi_{0,i}$$

but in this case the first order and second order corrections are more complicated. They are given by

$$\Delta W^{(1)} = \int \Psi_0^* \left(\frac{-qB}{2m} L_z \right) \Psi_0 \, d\tau$$

for the first-order correction to the energy of the ground-state wavefunction Ψ_0 and

$$\Delta W^{(2)} = \frac{q^2 B^2}{8m} \int \Psi_0^* (x^2 + y^2) \Psi_0 \, d\tau$$

$$\frac{-B^2 \sum_{\kappa > 0} \left(\Psi_\kappa^* \left(\frac{-qL_z}{2m} \right) \Psi_0 \, d\tau \right)^2}{W_\kappa - W_0}$$

The terms second order in the magnetic induction give the **magnetizability**, the magnetic analogue of polarizability. The first of these is known as the diamagnetic part, and it is very easy to calculate since it is just the expectation value of the second moment operators. The paramagnetic term is a very tricky customer, for a good reason that need not concern us in this text; The vector potential **A** and the magnetic induction **B** are related through the equation **B** = curl **A**, in the case of a time-independent **B**. Because of the vector identity curl (gradF) = 0 for any differentiable scalar field F, the vector potential is undefined to a sort of 'constant of integration'. Any calculated property must be independent of this constant of integration but, unfortunately, we have to work very hard indeed to satisfy the requirement.

Workers in the field talk about the problem of the 'choice of gauge', and there are relatively few calculations of magnetic properties generally in the literature.

15 Half a Dozen Applications

To finish off, I am going to describe briefly a few of my favourite applications. I hope that they will also interest you! Some are perennial but, more importantly, you will find that they make excellent research projects for senior undergraduates.

15.1 BARRIERS TO INTERNAL ROTATION AND INVERSION

The way in which information on the molecular structure of organic compounds is obtained has changed radically with time. The classical approach consisted of a chemical analysis leading to an empirical formula, but it eventually became obvious that molecular stereochemistries had to be evaluated when two or more substances with different chemical and physical properties were found to have the same constitution. The first concepts of stereochemistry were put forward by van't Hoff and Le Bel in the latter part of the nineteenth century. Van't Hoff postulated a tetrahedral model for the bonds associated with a carbon atom. He also postulated that free rotation could occur around a single C–C bond, but that rotation around multiple C–C bonds was restricted. As time passed, the concept of free rotation around single bonds became suspect, as indications were obtained that in many compounds free rotation was restricted and the molecule could exist in a number of possible stereoisomers.

Bischoff studied the constitution of a large number of organic substances and in a number of key papers he suggested that ethane in its equilibrium position has a staggered conformation, and that restricted rotation occurred in multiply substituted ethanes, such as the isomeric disubstituted succinic acids. The first experimental proof that restricted rotation could occur around single bonds came in 1922 when Christie and Kenner (1922) resolved 2,2'-dinitrophenyl-6-6'-dicarboxylic acid into optically active forms (Figure 15.1). Resolution is possible since there are two bulky groups on each ring, and these prevent rotation about the single middle bond. The barrier to internal rotation is high in this case.

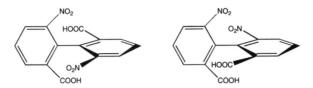

Figure 15.1   Optically active forms of 2,2'-dinitrophenyl-6-6'dicarboxylic acid

Figure 15.2   Barrier to rotation in ethane

Most experimental barriers in simple molecules have been determined from microwave spectroscopy. The best known value is 12.25 ± 0.11 kJ mol^{-1} for ethane, as determined by Weiss and Leroi (1968) (Figure 15.2). They fitted weakly allowed vibrational transitions to a torsional potential of the form

$$V - V_0 = V_3(1 - \cos 3\theta) + V_6(1 - \cos 6\theta) + \ldots$$

and concluded that only the V_3 term was necessary in order to achieve spectroscopic accuracy. Here, θ is the angle between CH groups on either C atom, as viewed down the CC axis. For molecules such as H_2O_2 and N_2H_4, etc. where the symmetry is lower, it is necessary to take more general forms of the potential such as

$$V - V_0 = V_1(1 - \cos \theta) + V_2(1 - \cos 2\theta) + V_3(1 - \cos 3\theta) + \ldots$$

and for asymmetric molecules it is necessary to add additional terms to reflect the lack of symmetry about 180°.

Let's go back to ethane. In order to calculate the V_1, $V_2 \ldots$ terms in the barrier, it is necessary to run electronic calculations for a range of values of θ and fit the results to the expression above. If terms V_2 and higher are negligible, then all we have to do is calculate the energies of the staggered and eclipsed conformations and subtract. I have done this in Table 15.1 for a three ZDO models and at the HF/4–31G level of theory. For the preliminary calculations, I left the geometries fixed with $R_{CC} = 153.6$ pm, $R_{CH} = 110.1$ pm and HCC = 108.4° All the models seem to perform reasonably, but from now on I will concentrate on the *Ab Initio* calculations.

Workers in the field usually emphasize the importance of geometry optimization, so I have done this at the HF/4-31G level, as shown in Table 15.2. We usually find that the eclipsed structure is slightly more 'open' than the staggered one, and this **reduces** the predicted barrier by a few per cent.

Most calculated properties are basis set dependent, so the next thing is to see how the calculated barrier depends on the sophistication of the basis set. This is shown in Table 15.3. There does not seem to be much to choose between basis sets on the basis of the calculations above. A more careful study of many molecules suggests that an extended basis set should normally be used.

Table 15.1 C_2H_6 barrier to rotation

| Model | HF energies/E_h | | Barrier/kJ mol^{-1} |
	Staggered	Eclipsed	
CNDO/2	−18.810939	−18.806947	10.48
MINDO3	−0.025221	−0.023154	5.43
AM1	−0.02544	−0.02318	5.93
HF/4−31G	−79.112115	−79.106190	15.56

Table 15.2   Geometry optimization

Parameter	Staggered	Eclipsed
R_{CC}/pm	152.9	154.1
R_{CH}/pm	108.3	108.3
HCC/°	111.2	111.
ϵ/E_h	−79.115933	−79.111519
	Barrier = 11.59 kJ mol^{-1}	

Table 15.3   Ethane barriers to internal rotation. Upper figures refer to the eclipsed form, lower figures to the staggered

Level	R_{CC} /pm	R_{CH} /pm	HCC/°	ϵ/E_h	Barrier /kJ mol^{-1}
HF/4G	154.6	108.2	111.1	−78.85862	12.00
	153.5	108.2	110.7	−78.86319	
HF/6−311G	154.0	108.2	111.7	−79.20736	11.67
	152.8	108.3	111.2	−79.21180	
HF/6−311G**	154.1	108.5	111.6	−79.24681	12.86
	152.7	108.6	111.2	-79.25171	
MP2/6−311G**		as above		−79.56558	13.40
as above				−79.57069	
experiment					12.25±0.11

I also ran a calculation at the MP2 level of theory, keeping the geometries fixed at the HF values. Electron correlation is thought to make a negligible contribution to rotational barriers about single bonds in substituted ethanes, but not for rotation about bonds such as the N–N bond in N_2O_4 or the central C–C bond in 1,3-butadiene.

The ammonia molecule is pyramidal, with an HNH bond angle of 106°. One mode of vibration can be visualized as that where the nitrogen atom moves to and fro through the plane of the hydrogen atoms. If the vibration is of sufficiently high energy, the nitrogen atom can reach a planar structure and then go through the plane. This process is known as **inversion**, and the energy difference between the planar and bent structures defines the **barrier to inversion**. The inversion barrier in ammonia is known to be 24.3 kJ mol^{-1} and on the basis of our discussion of barriers to internal rotation you might anticipate that an extended basis set HF calculation with geometry optimization at the bent and planar conformations would give good agreement with experiment.

Typical HF calculations are shown in Table 15.4. The first two calculations at HF/STO–4G and HF/4–31G levels of theory give very poor agreement with experimental results, and many early workers concluded that electron correlation was important. In fact, it is primarily a basis set effect, with electron correlation playing a very minor role.

This is a good example of the interplay between theory and experiment. The barriers have to be deduced from quite sophisticated experimental measurements, and careful attention has to be paid when choosing the correct level of theory. But in the end, the calculated values are of comparable accuracy to those produced experimentally, and so the theoretical model can be used to make predictions.

Table 15.4 Calculated geometries and barriers to inversion in ammonia: (b) = bent, (p) = planar

Level	R_{NH}/pm	HNH/°	ϵ/E_h	Barrier/kJ mol^{-1}
HF/4G (b)	102.9	104.4	−55.850214	43.86
(p)	100.3		−55.833509	
HF/4–31G	99.1	115.8	−56.106692	1.82
	98.6		−56.105997	
HF/6–311G**	100.1	107.4	−56.210397	23.63
	98.7		−56.201398	
MP2/6–311G**	101.4	106.0	−56.408790	27.01
	99.7		−56.398501	
CISD/6–311G**	101.2	106.1	−56.414195	27.51
	99.5		−56.403718	

15.2 HYDROGEN-BONDED COMPLEXES

When a covalently bound hydrogen atom forms a second bond to another atom, this second bond is referred to as a hydrogen bond. These bonds can be intermolecular or intramolecular and in the species A–H . . . B, A is usually an atom of electronegativity greater than hydrogen Credit for the discovery of hydrogen bonds is usually given to Latimer and Rodebush (1920) and we have, obviously, come on a long way since they wrote the water dimer as depicted in Figure 15.3.

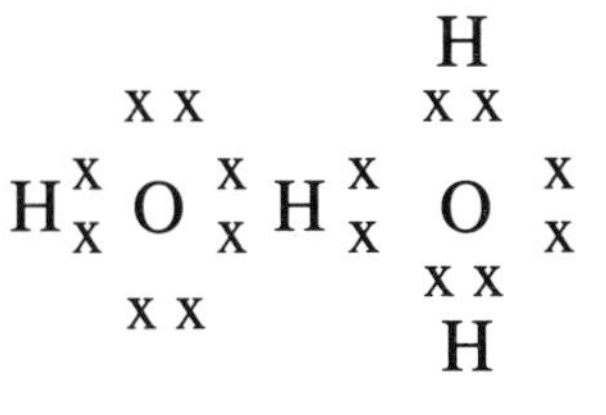

Figure 15.3 1920s view of the water dimer

The first *Ab Initio* molecular orbital study seems to be that of Clementi (1961) on HF_2^-. In the 1980s, new techniques in spectroscopy such as molecular beam electric resonance spectroscopy and pulsed nozzle Fourier transform microwave spectroscopy were developed. These sophisticated experiments gave for the first time accurate molecular information on small hydrogen-bonded complexes. A large number of complexes have been studied and a selection is shown in Figure 15.4. In the cyclopropane complex, the HCl bonds to the area of high electron density associated with the region outside the ring. The HCl molecule lies along a C_2 symmetry axis. In the benzene case, the HF bonds to the π-electron density, and the complex has a C_6 symmetry axis.

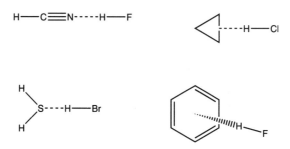

Figure 15.4   Some simple hydrogen-bonded complexes

For illustration, let's take HCN...HF, which turns out experimentally as a linear complex. The simplest way to tackle the problem within the LCAO method is to calculate individually the energies of HCN and HF (the 'monomers'), and then calculate the energy of the complex (the 'dimer'). This gives ΔU for the reaction

$$HCN(g) + HF(g) \rightarrow HCN...HF(g)$$

which is usually referred to as the hydrogen bond energy. We refer to this technique as the **supermolecule** method and Table 15.15 records results at the HF/4–31G level of theory. I optimized the geometries in order to calculate the harmonic force constants. Hydrogen-bonded complexes are relatively weakly bound, so to a first approximation we could have kept the 'monomer' geometries constant for the 'dimer' calculation.

Table 15.5   HCN...HF complex with frozen monomers, HF/4–31G level

Molecule	Bond	Length/pm	ϵ/E_h
HCN	HC	114.0	−92.731929
	CN	105.1	
HF	HF	92.2	−99.887287
HCN...HF	HC	105.2	−192.633841
	CN	113.7	
	HF	92.9	
	N...H	190.3	

The energy difference is $38.40\,\text{kJ}\,\text{mol}^{-1}$, compared with the experimental D_e of $26.1 \pm 1.6\,\text{kJ}\,\text{mol}^{-1}$. The experimental $N\ldots F$ distance is 279.2 pm, compared with the calculated value of 283.2 pm. The primary experimental quantity is $D_0 = 18.9 \pm 1.1\,\text{kJmol}^{-1}$ but this was corrected to the D_e value given by assuming harmonic vibrations. We can investigate the zero point contribution by calculating harmonic force constants at the energy minima. These zero point contributions are 24.63, 47.17 and $80.30\,\text{kJ}\,\text{mol}^{-1}$, giving a calculated D_0 of $29.90\,\text{kJ}\,\text{mol}^{-1}$. The hydrogen bond energy is basis set dependent, and a recalculation at the HF/6–311G** level of theory gives $D_e = 27.48\,\text{kJ}\,\text{mol}^{-1}$, in excellent agreement with experiment. The energetics and properties of complexes held together with moderately strong hydrogen bonds are well represented by SCF calculations with sensible basis sets.

The complex formed between CO and HF is interesting because the hydrogen bond is weak. The complex turns out to be linear, and there are two possibilities; $CO\ldots HF$ and $OC\ldots HF$. The simplest theories of hydrogen bonding concentrate on multipole–multipole interactions, so we might expect that the $OC\ldots HF$ arrangement would have the lower energy of the two. On the basis of our previous discussion about the sign of the dipole moment in HF, we might expect to find some difficulties with Hartree Fock supermolecule calculations. Table 15.6 shows HF/6–311G** calculations for CO, HF, $CO\ldots HF$ and $OC\ldots HF$.

Table 15.6 $CO\ldots HF$ and $OC\ldots HF$ complexes at the HF/6–311G** level of theory

Molecule	Bond	Length/pm	ϵ/E_h
CO		110.5	-112.769475
HF		89.6	-100.046904
$CO\ldots HF$	CO	110.7	-212.819624
	$O\ldots H$	216.1	
	HF	89.7	
$OC\ldots HF$	CO	110.2	-212.820018
	$C\ldots H$	233.1	
	HF	89.8	

This gives hydrogen bond values of D_e of $8.52\,\text{kJ}\,\text{mol}^{-1}$ and $9.55\,\text{kJ}\,\text{mol}^{-1}$ for the two $CO\ldots HF$ and $OC\ldots HF$ possibilities. It is interesting to note that the $CO\ldots HF$ dimer and the $OC\ldots HF$ dimer have roughly the same stability at HF level of theory, despite the incorrect behaviour of the CO dipole moment.

The argument goes that electron correlation is almost certainly of importance in such complexes where the binding energy is small, so I have repeated the calculation at the MP2/6–311G** level of theory (Table 15.7). The hydrogen bond energies now become 7.57 and $15.68\,\text{kJ}\,\text{mol}^{-1}$. The conclusion is that electron correlation has to be treated when considering weakly bound complexes.

There are several interesting aspects to these calculations. First of all, the complexes are very loosely bound and have very small bending force constants (they are described as 'floppy'). Secondly, there is a problem when attempting to describe weak interactions within the supermolecule LCAO model, where we calculate the energy difference between A–H...B and A–H and B. Suppose that the basis set employed is small. The orbitals of the A–H monomer will 'borrow' basis functions belonging to B in the complex, and the orbitals of B will also borrow basis functions belonging to A–H. Small basis sets usually give hydrogen bond energies that are much too large, for this reason.

Table 15.7   CO...HF and OC...HF complexes at the MP2/6–311G** levels of theory

Molecule	Bond	Length/pm	ϵ/E_h
CO		113.9	−113.074756
HF		91.3	−100.267216
CO...HF	CO	114.0	−213.344856
	O...H	213.2	
	HF	91.3	
OC...HF	CO	113.7	−213.347946
	C...H	214.8	
	HF	91.7	

15.2.1   Ghost orbitals

Boys and Bernardi (1970) have proposed a scheme called the **counterpoise method** whereby this so-called basis set superposition error can be compensated for. One calculates the interaction energy for the complex A–H...B as the difference

$$U(A–H...B) - U'(A–H) - U'(B)$$

where $U(A–H)$ is the energy of the monomer A–H calculated in the presence of the orbitals centred on monomer B (but without the nuclei comprising B). Likewise $U'(B)$. These orbitals are often referred to as **ghost orbitals**, and I have illustrated their use in Table 15.8 for calculations on HCN...HF. There have been many reviews on hydrogen-bonded complexes; Hasenein and Hinchliffe (1990) is just one example.

Table 15.8   Use of ghost orbitals for HCN...HF complex at the SCF level of theory

Basis set	R(N...H) /pm	R(H–F) /pm	$D_e/kJ\,mol^{-1}$	
			ΔSCF	Ghost
STO–4G	207.9	95.5	15.3	8.3
4-31G	192.7	92.2	38.8	32.6
Double	186.3	91.9	22.6	35.2
Extended, polarized	197.2	90.5	27.5	24.7
Experiment	$R_0(N...F) = 279.2\,pm$ $D_e = 26.1 \pm 1.6\,kJ\,mol^{-1}$			

15.3 THE SPHERICAL FLOATING GAUSSIAN (FSGO) MODEL

In standard quantum mechanical molecular structure calculations, we choose a set of basis functions $\chi_1, \chi_2 \ldots \chi_n$ that are nuclear centred. These functions are often chosen with a view to the ease of integral evaluation, which explains the popularity of GTOs. Once the basis functions have been chosen, we would proceed with an SCF or a VB calculation, or whatever.

The simplest possible many-electron wavefunction with the necessary antisymmetry is represented by a single Slater determinant. Since most molecules have an even number of electrons, and have singlet electronic ground states, the Slater determinant can be formed from doubly occupied orbitals. In Chapter 5 I wrote

$$\Psi_{el}(\underline{x}_1, \underline{x}_2 \ldots \underline{x}_{2M}) = N \begin{vmatrix} \psi_A(\mathbf{r}_1)\alpha(s_1) & \psi_A(\mathbf{r}_1)\beta(s_1) & \cdots & \psi_M(\mathbf{r}_1)\beta(s_1) \\ & \cdots & & \\ \psi_A(\mathbf{r}_{2M})\alpha(s_{2M}) & \psi_A(\mathbf{r}_{2M})\beta(s_{2M}) & \cdots & \psi_M(\mathbf{r}_{2M})\beta(s_{2M}) \end{vmatrix}$$

where there were M orbitals ψ_A through ψ_M, and each was doubly occupied. The factor N is the normalizing factor. In the case of orthonormal values of ψ, N is given by

$$N = \frac{1}{\sqrt{(2M)!}}$$

In the case of traditional LCAO approaches, the ψ's are formed as linear combinations of the basis functions $\chi_1, \chi_2 \ldots$ and the LCAO coefficients are found by solving the matrix SCF equations of an earlier Chapter 5.

The floating spherical Gaussian orbital (FSGO) model in its simplest form applies only to such systems. The many-electron wavefunction is taken to be a single Slater determinant formed by double occupancy of single spherical (1s-type) Gaussian orbitals centred at arbitrary positions within the molecule. The FSGOs are allowed to move freely through space (to 'float'), and to vary in size (by changing the orbital exponent α) until the variational energy is a minimum.

In the FSGO model, we just consider five spherical Gaussians for H_2O (which has ten electrons). Because each FSGO is doubly occupied, there is no SCF calculation! No iterations are required, and the variational energy can be easily calculated. So, for $2M$ electrons we choose M Gaussian orbitals $\chi_1, \chi_2 \ldots \chi_M$ and the Slater determinant is now

$$\Psi_{el}(\underline{x}_1, \underline{x}_2, \ldots \underline{x}_{2M}) = \begin{vmatrix} \chi_A(\mathbf{r}_1)\alpha(s_1) & \chi_A(\mathbf{r}_1)\beta(s_1) & \cdots & \chi_M(\mathbf{r}_1)\beta(s_1) \\ & \cdots & & \\ \chi_A(\mathbf{r}_{2M})\alpha(s_{2M}) & \chi_A(\mathbf{r}_{2M})\beta(s_{2M}) & \cdots & \chi_M(\mathbf{r}_{2M})\beta(s_{2M}) \end{vmatrix}$$

The normalizing factor is different because the basis functions overlap each other. In terms of the matrix of overlap integrals $\underline{S}$

$$N = \frac{1}{\sqrt{(2M)!}\ \det \underline{\mathbf{S}}}$$

where $\det \underline{\mathbf{S}}$ is the determinant of $\underline{\mathbf{S}}$, whose elements $(\underline{\mathbf{S}})_{ij}$ are $\int \chi_i(\mathbf{r})\chi_j(\mathbf{r})d\tau$.

The energy formula is similar to that derived for the corresponding SCF method, except that the orbitals are not orthogonal. The electronic energy expression turns out to be

$$\epsilon_{\mathrm{el}} = 2 \sum_{R=A}^{M} \sum_{S=A}^{M} T_{RS} \int \chi_R(\mathbf{r}_1) h(\mathbf{r}_1) \chi_S(\mathbf{r}_1) d\underline{\tau}_1$$

$$+ \sum_{R=A}^{M} \sum_{S=A}^{M} \sum_{T=A}^{M} \sum_{U=A}^{M} \int\int \chi_R(\mathbf{r}_1)\chi_R(\mathbf{r}_1) g(\mathbf{r}_1, \mathbf{r}_2)\chi_S(\mathbf{r}_2)\chi_S(\mathbf{r}_2) d\underline{\tau}_1 d\underline{\tau}_2$$

$$\times (2T_{RS}T_{TU} - T_{RU}T_{ST})$$

where the matrix $\underline{\mathbf{T}} = \underline{\mathbf{S}}^{-1}$. The calculation proceeds by choosing a set of FSGOs, each of which is characterized by its position in space and its size (the Gaussian exponent). The overlap integrals are then calculated, and the overlap matrix $\underline{\mathbf{S}}$ inverted. The electronic energy is then calculated; the formulae for one and two-electron integrals over 1s Gaussians are quite simple and were, of course, first given by Boys. After an energy evaluation, the parameters are varied until a minimum energy is reached. If a geometry optimization is required, then one varies the nuclear positions also, as an 'outer loop' to the calculation.

Thus, in the simple case of LiH, we would expect to find two FSGOs; a small one near the Li nucleus, and a larger one centred along the internuclear axis to represent the bonding. The results are given in Table 15.9 and are illustrated in Figure 15.5. The total energy is only about 80% of comparable HF/4–31G value. The predicted bond length in this particular example is about 8% larger than experimental results.

Table 15.9   FSGO calculation on LiH

Li nuclear coordinate/a_o	0.0000
H nuclear coordinate/a_o	3.2260
Inner shell coordinate/a_o	−0.0076
Inner shell exponent	1.4136
Bonding orbital coordinate/a_o	2.8829
Outer shell exponent	0.4107

It turns out that FSGO energies are almost always 80–85% of the corresponding LCAO values, but the errors in the predicted bond lenths are usually quite respectable. In the special case of LiH, the bonding orbital is centred close close to the H nucleus, which reflects the highly ionic nature of the molecule. Despite the crudity of the FSGO model, bond lengths are given to good accuracy. Molecular conformations are also well represented. The first applications of the FSGO model were reported by Frost (1967, 1968). Suitable computer programs are discussed in Frost et al (1968).

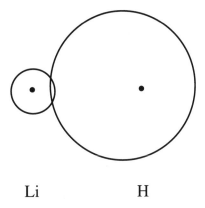

Li   H

Figure 15.5   LiH FSGO model

All seems to be quiet on the FSGO front at the minute, but you might like to know of Christofferson's work (1971). Their idea was to build up large molecules out of small fragments, using the FSGO calculations on these fragments to model the target molecule. For example, to form ethane we start from two 'fragments' and subtract a H atom from either. The two vacant bonding C–H FSGOs are then taken to represent the C–C bond orbital in ethane (plus, of course, a virtual orbital). Some peptide molecules have been studied by this procedure.

15.4   ISOTROPIC HYPERFINE COUPLING CONSTANTS

I wrote my PhD thesis in 1969, and the Introduction went thus.

'Up to about 10 years ago, the main theoretical interest in organic ions and radicals was the calculation of their ionization energies and electron affinities. Some theoretical interest was also shown in their electronic spectra.

With the advent of magnetic resonance techniques, and in particular electron spin resonance, it became possible to study experimentally more subtle features of molecular structure. For example, couplings between electrons which lead to zero field splittings in triplet states, and couplings between electrons and nuclei which lead to the isotropic proton hyperfine structures observed in ESR spectra of organic ions and radicals.

The theory of these magnetic interactions was reported by Fermi as early as 1930 (Fermi 1930), and the numerical calculation of the parameters describing these properties gives a testing ground for theoretical methods. In addition, it gives useful information about the electronic structure of large radicals, triplet states and ions.'

In an earlier Chapter, I gave a brief indication of the quantum mechanical treatment of a **single** electron in an external electromagnetic field. The field-free Hamiltonian for a free electron

$$H = \frac{1}{2m}p^2$$

is modified to

$$H = \frac{1}{2m}(\underline{\mathbf{p}} - e\underline{\mathbf{A}})^2 - e\phi$$

where ϕ is the scalar potential and $\underline{\mathbf{A}}$ the vector potential.

I have to tell you that this treatment is only half right and the reason is connected, amongst other things, with my rather cavalier treatment of spin. The concept of electron spin is usually introduced in a 'phenomenological' way, without mathematical justification. We simply note that the Stern Gerlach experiment demonstrated that electrons have an intrinsic angular momentum. Total electron wavefunctions are, therefore, written as spatial parts (derived from solutions of the Schrödinger equation), and another quantity that we refer to as 'spin'. Many other particles possess spin, and the interesting thing is that spin is an angular momentum with half-integral quantum numbers. Spin certainly does **not** arise from the Schrödinger treatment. Dirac (1928) replaced the equation above by its relativistic counterpart

$$H = [m^2c^4 + (\underline{\mathbf{p}} - e\underline{\mathbf{A}})^2]^{\frac{1}{2}} - e\phi$$

and found that the concept of electron spin appeared naturally from his treatment.

For a many particle system, the situation is much more uncertain. Breit (1930) set up a two particle analogue of the Dirac equation, and the resulting Hamiltonian contains all terms that one might expect (such as the dipole–dipole interaction), plus many more non-classical ones. If we follow the usual approximate treatment discussed by McWeeny and Sutcliffe, then the molecular Hamiltonian has to be modified by adding the following perturbations H';

$$H' = H_{\text{elec}} + H_{\text{mag}} + H_{\text{SL}} + H_Z + H_{\text{SS}} + H_N$$

where H_{elec} arises from external electric fields; H_{mag} arises from external magnetic fields interacting with electronic orbital motion; H_{SL} arises from interactions between electron spin and orbital motion: H_Z arises from Zeeman interactions between the electron spin and the magnetic field; H_{SS} arises from electron spin–spin interactions; and H_N contains all the so-called hyperfine terms arising from nuclear spins.

In the case of the electron spun resonance (ESR) experiment, the only terms of interest are the following contributions to H_N

$$g\beta \sum_{\alpha=1}^{\text{NUC}} \sum_{i=1}^{n} g_\alpha \beta_\alpha \left\{ \frac{[3\underline{\mathbf{s}}(i).\underline{\mathbf{r}}_{\alpha,i}]\,[\underline{\mathbf{I}}(\alpha).\underline{\mathbf{r}}_{\alpha,i}]}{r_{\alpha,i}^5} - r_{\alpha,i}^2 \underline{\mathbf{I}}(\alpha).\underline{\mathbf{s}}(i) \right\}$$

and

$$\text{g}\beta \sum_{\alpha=1}^{\text{NUC}} \sum_{i=1} \frac{8\pi}{3} \delta(\mathbf{r}_{\alpha,i})\underline{\mathbf{I}}(\alpha).\underline{\mathbf{s}}(i)$$

The first term gives the classical dipole interaction between the nuclear spin dipoles and the electron spin dipoles. In solution, and in the gas phase, it averages to zero. The second term is the so-called *contact interaction*. The Dirac delta function $\delta(\mathbf{r}_\alpha, i)$ is zero unless electron i is at the position of nucleus α.

ESR spectroscopists interpret their spectra in terms of a **spin Hamiltonian**, which they would write

$$H_{\text{spin}} = h \sum_{\alpha=1}^{\text{NUC}} a_\alpha S_z(i) I_{\alpha,z}(\alpha)$$

The idea is that H_{spin} only operates on spin states (both electron and nuclear), and the details of the electronic motions are averaged into the **hyperfine coupling constant** a_α for nucleus α. Analysis of the spectrum yields magnitudes of the a, but not their sign. The signs of the values of a **can** be deduced from single crystal studies, but many of the free radicals and triple states under study are often very unstable.

Simple quantum mechanical calculations give the magnitude and the sign of the values of a, and from the 1960s, the literature has afforded many examples of predictive calculations.

The relationship between a_α and the electronic wavefunction is derived in all the specialist ESR texts as

$$a_N = \frac{4\pi}{3} g\beta g_N \beta_N \frac{h}{\langle S_z \rangle} Q_1(\mathbf{R}_\alpha)$$

where $Q_1(\mathbf{R}_\alpha)$ is the spin density function of Chapter 4, evaluated at the position of nucleus .

A simple calculation on the H atom will make this clear. The exact atomic orbital is a STO,

$$\chi_{1s}(\mathbf{r}) = \sqrt{\frac{1}{\pi a_0^3}} \exp(-r/a_0)$$

and so the spin density evaluated at the nucleus is $1/(\pi a_0^3)$. This gives a coupling constant of 50.765 mT, which compares very well with the experimental value of 50.682 mT. The difference is connected with the derivation of the formula for the spin coupling constant.

Hyperfine coupling constants (HFCC) depend on the behaviour of the spin density, evaluated at the position of the nucleus in question. Molecular *Ab Initio* calculations invariably use GTOs, and I mentioned in Chapter 8 that GTOs have the wrong behaviour at their nuclear centre. Consequently, we have to work much harder with

GTO expansions, in order to arrive at anything like the correct HFCC even for a H atom. Table 15.10 says it all

Table 15.10   Hyperfine coupling constants in the H atom

Level of theory	a_H/mT
STO	50.765
GTO; STO–3G	63.005
STO/6G	85.532
4–31G	47.554
6–311G*	45.884
Experiment	50.682

Molecular calculations are approached in the usual way:

- choose the level of sophistication of the calculation;
- choose the basis set;
- decide whether a geometry optimization is necessary; and then
- run the package.

An example should suffice.

The allyl radical CH_2CHCH_2 (Figure 15.6) is a simple π-electron system, and the odd electron is in a π orbital.

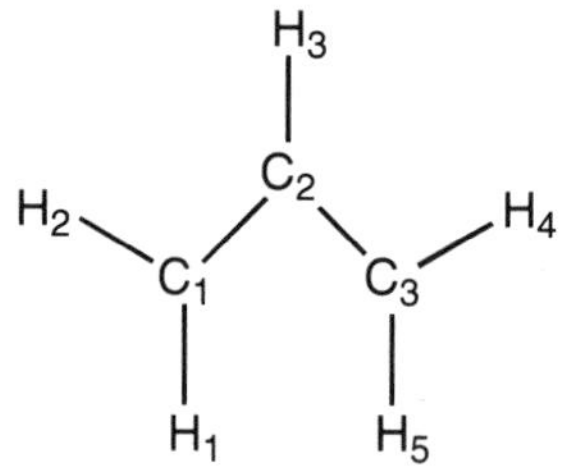

Figure 15.6   The allyl radical

The radical has been the subject of many theoretical and experimental studies since the 1960s, and the proton coupling constants have been derived from single crystal studies on related molecules as $+1.483$, $+1.393$ for the CH_2 protons and $-0.406\,mT$ for the CH proton. Formally, the odd electron resides in a orbital, and as the radical is planar, the spin density at each nuclear position is predicted to be zero. This was explained by McConnell in terms of the **spin polarization** mechanism; the spin due to the odd electron polarizes the electron spins in each CH bond, and so the H nuclear spin can sense the electron spin at the C atom, through the CH -electrons. Pictorially, the argument runs as shown in Figure 15.7.

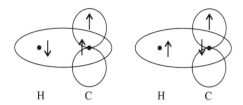

Figure 15.7   The spin polarization mechanism

The two electrons in the CH bond will be spin paired, and are thought of as being 'localized' at either end of the CH bond. Think about the two possibilities as valence bond structures, and the left hand structure will be slightly more preferred, depending on the magnitude of the spin density on the C atom. McConnell's formula is

$$a_N(\text{proton}) = -2.3\,mT \times C\,\text{spin density}$$

A Hückel π-electron calculation gives the odd electron MO as

$$\psi_B = \sqrt{\frac{1}{2}}(\chi_1 - \chi_3)$$

and so the carbon spin densities are taken to be 1/2 ,0 and 1/2. In particular, the 'middle' proton C_3 is predicted to have a zero HFCC, in contrast to the experimental value. (Note that the 'spin densities' used here are just the elements of the spin density matrix. They are not the values calculated at the nuclear positions).

At the *Ab Initio* level, we don't invoke the McConnell (1956) relationship, but calculate the proton spin densities exactly. Even so, at ROHF levels of theory, the odd electron occupies an orbital which has a nodal plane that includes the CH group, and the HF spin densities come out as zero. The lowest level of *Ab Initio* HF calculations needed for such radicals is UHF, where the spin density is given as the difference between the density of the spin electrons minus that due to the spin ones. The spin density function then has to be evaluated at the nuclear positions. Alternatively, we have to make ROHF calculations but then allow for electron correlation. *Ab initio* UHF calculations are shown in Table 15.11. I restricted the molecule initially to C_{2v} symmetry, but optimized the geometry in each case.

Table 15.11   Proton hyperfine coupling constants for allyl radical

Level of theory	$<S^2>$	a_1/mT	a_2/mT	a_3/mT
UHF/STO–4G	1.112	−5.26	−5.35	+4.13
UHF/4–31G	0.984	−4.23	−4.33	+2.95
UHF/6–311G**	0.965	−3.23	−3.34	+2.16
Experiment	3/4	−1.39	−1.48	+0.41

At first sight, the agreement with experimental results is appalling. The reason is that whilst UHF wavefunctions are automatically eigenfunctions of the S_z spin operator, they are **not** eigenfunctions of S^2. This is shown in the table as the expectation value $<S^2>$, and the interpretation is that a value very far from the true doublet value of $3/4$ means serious spin contamination from an electronic state of higher spin multiplicity. For example, the true doublet electronic state of a simple radical under study might be planar, but a nearby quartet state could be distinctly non-planar. The resultant ground state UHF treatment of the radical would predict non-planarity because of spin contamination from the quartet state.

Working on the assumption that the major contamination in the doublet UHF wavefunction is due to a quartet state, Amos and Snyder (1965) showed how to eliminate the quartet state from the UHF wavefunction. It is usually observed that this procedure reduces substantially the HFCC.

Table 15.12 shows results from correlated *Ab Initio* calculations. In all cases, the 4–31G basis set was used, and the geometry was fixed at the HF/4–31G level.

Table 15.12   Proton hyperfine coupling constants for allyl radical (4–31G basis set)

Level of theory	a_1/mT	a_2/mT	a_3/mT
MP2	−2.51	−2.60	+0.70
CID	−4.08	−4.17	+2.85
CISD	−3.37	−3.46	+2.13
Experiment	−1.39	−1.48	+0.41

The agreement with experimental results leaves a lot to be desired, and the best advice that I can give is that you should go back to Hückel or PPP levels of theory!

15.5   ISODESMIC REACTIONS

Many people would argue that the 'stuff' of chemistry is reactions. Thus, successful models of chemistry should be able to calculate the thermodynamic quantities associated with a given reaction (the internal energy change and the entropy change) and give information about the reaction mechanism.

Let's start with the simpler of those, and concentrate for the minute on thermodynamic properties.

This seems a simple goal, but it proved elusive in the early days of molecular modelling. The problem is that of electron correlation, and I have drawn your attention to this on several occasions in the text. Even the simple reaction

$$H_2 \rightarrow 2H(^2s)$$

is beyond the scope of HF theory, because of the electron correlation error. The HF wavefunction dissociates into **ions** rather than **atoms**

$$H_2 \rightarrow H^- + H^+$$

and so the 'predicted' HF internal energy change is completely in error.

Some authors have been more modest in their goals and attempted to calculate directly the internal energies involving closed shell species, where there is some prospect that the correlation error will at least cancel. The two papers normally quoted in this field are Snyder & Basch (L C Snyder and H Basch, J Amer Chem Soc 91 (1969) 2189), and the following.

Molecular Orbital Theory of the Electronic Structure of Organic Compounds V Molecular Theory of Bond Separation
W. J. Hehre, R. Ditchfield, L. Radom and J. A. Pople
Journal of the American Chemical Society **92** (1970) 4796

The complete hydrogenation of an organic molecule is separated into two processes. In the first, termed **bond separation**, the molecule is separated into its simplest parents containing the same component bonds. The energy associated with such a reaction is then the **heat of bond separation.** The second step then consists of full hydrogenation of the products of bond separation. To study these two processes, we have performed *Ab Initio* molecular orbital calculations on a variety of polyatomic molecules. Both minimal and extended basis sets, taken as linear combinations of Gaussian-type functions, are shown to give heats of bond separation in good agreement with experiment. In contrast, only the extended basis is successful in reproducing the heats of hydrogenation of the parents.

Let it pass that the authors do not seem to care about the distinction between 'heat', 'internal energy' and 'enthalpy'.

Thus, for example, consider benzene. The idea is to predict the enthalpy change for a reaction such as

$$C_6H_6 + 9H_2 \rightarrow 6CH_4$$

We use Hess' Law to split up the reaction as follows

$$C_6H_6 + 6CH_4 \rightarrow 3C_2H_6 + 3C_2H_4 \tag{1}$$

and

$$C_2H_6 + H_2 \rightarrow 2CH_4$$
$$C_2H_4 + 2H_2 \rightarrow 2CH_4 \tag{2}$$

Addition of the two sets of equations then gives the master equation for complete hydrogenation

$$C_6H_6 + 9H_2 \rightarrow 6CH_4$$

Bond separation reactions such as (1) are examples of chemical changes in which there is retention of the number of bonds of a given formal type, but with a change in their relationship to one another. Such equations are called **isodesmic**, and we could predict the internal energy of the molecule knowing the internal energy change for the reaction, together with the internal energies of the other reactants and products.

To see how it fits together, consider the data in Table 15.3. I ran HF/6–311G** OPT calculations on the molecules, and also calculated their vibrational frequencies at the equilibrium geometries.

Table 15.13 Molecular and thermodynamic data, calculated at the HF/6–311G** level

Molecule	ϵ_{el}/E_h	Corrected to 298 K
H_2	−1.132491	−1.119664
CH_4	−40.209012	−40.159212
C_2H_4	−78.054725	−77.997729
C_2H_6	−79.251708	−79.169691
C_6H_6	−230.754098	−230.643386

Some care is needed when comparing the results of such calculations with thermodynamic data. First of all, is the experimental data to do with **internal energy** (*U*) or **enthalpy** (*H*)? The two are related, for an ideal gas reaction by the well-known equation

$$\Delta H = \Delta U + RT\Delta n$$

where Δn is the change in the number of gaseous moles. Theoretical data is almost certainly internal energies. Secondly, experimental data sometimes refers to D_0 and sometimes to D_e. Even for a diatomic molecule at 0 K, there is a zero-point contribution to the vibrational energy of

$$D_e = D_0 + \frac{1}{2}hc_0\omega_e$$

and this is why I calculated the vibrational frequencies, and explains the Table entries 'corrected to 298 K' in the table. Like is (I believe) compared to like in Table 115.14.

Table 15.14 Internal energy changes at 298 K and 1 atm pressure

Reaction	Calculated/kJ mol^{-1}	Experiment/kJ mol^{-1}
$CH_3–CH_3 + H_2 \rightarrow 2CH_4$	−76.17	−75.73
$CH_2=CH_2 + 2H_2 \rightarrow 2CH_4$	−213.63	−194.56
$C_6H_6 + 6CH_4 \rightarrow 3CH_2=CH_2$		
$+ 3CH_3–CH_3$	+253.09	+275.30

The agreement with experiment is very good, for both types of reaction. In this particular case (benzene), the bond separation energy refers to the energy of the molecule compared to single and double bonds, and so the energy change should be a good estimate of the resonance energy.

15.6   CRYSTALLINE LATTICES SUCH AS ZEOLITES

Zeolites are inorganic materials that participate in many acid-catalysed industrial processes. It is found experimentally that the structures and chemical properties of these compounds are mainly determined by the relative numbers of the silicon to aluminium atoms. These Al and Si atoms occur in the zeolite framework as units like SiO_4^{4-} and AlO_4^{5-}, and they are linked between themselves through a bridging hydroxyl group called the 'Brönsted acid site'.

Structurally, these Brönsted acid sites are located at the surface of zeolite networks and are thought to be the main species responsible for the considerable catalytic powers of these materials.

Many zeolite acid-catalysed reactions are thought to occur by proton transfer between the acid site, and these adsorption processes lead to complexes of varying strength. They are classified as 'ionic', 'hydrogen-bonded' or 'van der Waals' depending on the strength of the interaction. Zeolite modelling is topical.

The idea is to model an infinite lattice as a very small 'cluster'. The composition of the cluster is open to argument, as is the size of this cluster. Naturally, workers want to take the smallest possible cluster at the *Ab Initio* level, or the largest cluster possible within standard semi-empirical methodology. You might like to read the following typical *Ab Initio* paper, which concerns H_2S and its complexes with zeolite clusters.

An *Ab Initio* Study of the Molecular Complexes Formed
between H_2S and the Acid Sites of Zeolites
Humberto J. Soscún M, Patrick J. O'Malley and Alan Hinchliffe
Journal of Molecular Structure (THEOCHEM) paper in Press 1995

Molecular complexes formed by the interaction of hydrogen sulfide H_2S with silanol $H_3Si(OH)$, two model Brönsted acid sites of zeolite clusters $H_3Si(O-H)AlH_3$ and $H_3Si(OH)Al(OH)_2SiH_3$ and a sodium zeolite cluster $H_3Si(O-Na)Al(OH)_2SiH_3$, have been studied by *Ab Initio* methods. Both Hartree Fock (HF) and post Hartree-Fock (MP2) levels of theory were used. The geometries of the isolated molecules and the corresponding complexes were optimized at HF/6–31G** level of theory, assuming C_s symmetry for the complexes.

Corresponding geometry optimization at the MP2/6–31G** level of theory were also carried out for the H_2S-silanol and the H_2S-$H_3Si(OH)AlH_3$ complexes.

Comparisons with similar water–zeolite complexes are made. The HF harmonic vibrational frequencies of the OH and SH stretching modes of these complexes are compared with experimental infrared measurements.

We took three models for H_2S adsorption on the base B. Figure 15.8 is the simplest, where H_2S interacts with model base B_0 and figure 15.9 illustrates our most complex.

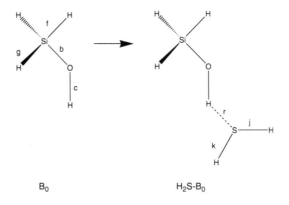

Figure 15.8   Effect of H_2S on B_0

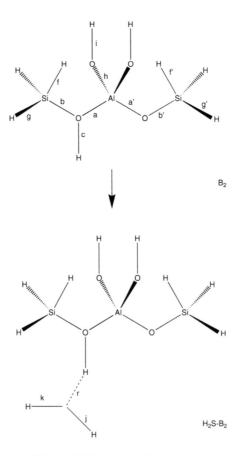

Figure 15.9   Effect of H_2S on B_2

It is usual to try and model infinite lattices using simple clusters of the complexities shown.

We concluded as follows.

'The molecular structures compare well with experimental findings, which suggests the formation of a linear, a cyclic and a coordinative complex structure for the interaction between H_2S and . . . the acid sites in zeolites . . .'

The size of the cluster used to model the OH sites in zeolites is important . . .'

The binding energy for the complexes is much less than that corresponding to the silanol. . .water complex, indicating that the strength of the H_2S . . . zeolite interaction is less than the water. . .zeolite value. This is consistent with experimental observations of the higher acidity of H_2S in comparison with water.'

References

Ahmed, F. and Cruickshank, D. W. J. (1952) *Acta Crystallographica*, **5**, 852.

Allinger, N. L., Yuh, Y. H. and Lii, J-H. (1989) *J. Amer. Chem. Soc.*, **111**, 8566, 8576.

Amos, A. T and Snyder, L. C. (1965) *J. Chem. Phys.*, **42**, 3670.

Andrews, D. H. (1930) *Phys. Rev.* **36**, 544.

Bates, D. R., Ledsham, K. and Stewart A. L. (1953) *Phil. Trans. Roy. Soc.*, **A246**, 215.

Bingham, R. C., Dewar, M. J. S. and Lo, D. H. (1975) *J. Amer. Chem. Soc.*, **97**, 1285–1293.

Bogaard, M. P. and Orr, B. J. (1975) *Int. Rev. Science Phys. Chem. Series 2*, Vol. 1, 149, Butterworths.

Born, M. and Oppenheimer, R. (1927) *Ann. Phys.*, **84**, 457.

Boys, S. F. (1950) *Proc. Roy. Soc. Ser A*, **200**, 542–544.

Boys, S. F. and Bernardi, F. (1970) *Molec. Phys.*, **19** 558.

Boys, S. F. and Cook, G. B. (1960) *Reviews of Modern Physics*, **32**, 285–295.

Breit, G. (1930) *Phys. Rev.*, **35**, 1447.

Brillouin, L. (1933) *Actual Sci. Ind.*, **71**.

Brooks, B. R. et al. (1983) *J. Comput. Chem.*, **4** 187.

Burkert, U. and Allinger, N. L. (1982) *Molecular Mechanics*, American Chemical Society, Washington DC.

Burrau, 210. K. (1927) *Danske Videnske Selsk. Mat-Fys Medd.*, **7**, 1.

Burrus, C. A. (1958) *J. Chem. Phys.*, **28**, 427.

Christie, G. H. and Kenner, J. (1922) *J. Chem. Soc.*, **LXXI**, 614.

Christofferson, R. E, Genson D. W. and Magiroa G. M. (1971) *J. Chem. Phys.*, **54**, 239.

Clementi, E. (1973) *Proceedings of the Robert A Welch Foundation Conferences on Chemical Research XVI, Theoretical Chemistry* (Robert A. Welch Foundation, Houston).

Clementi, E. (1961) *J. Chem. Phys.*, **34**, 1468.

Clementi, E. (1964) *J. Chem. Phys*, **40**, 1944–1945.

Clementi, E. and Raimondi, D. L. (1963) *J. Chem. Phys.*, **38**, 2686–2689.

Collins, J. B., Schleyer, P. von R., Binkley, J. S. and Pople, J. A. (1976) *J. Chem. Phys.*, **64**, 5142–5151.

Del Bene, J. and Jaffé, H. H. (1968) *J. Chem. Phys.*, **48**, 1807–1813.

Dewar, M. J. S. and Walter, T. (1977) *J. Amer. Chem. Soc.*, **99**, 4907–4917.

Dickinson, B. (1933) *J. Chem. Phys.*, **1**, 317.

Dirac, P. A. M. (1928) *Proc. Roy. Soc. (London) Series A*, **117**, 610.

Dirac, P. A. M. (1929) *Proc. Roy. Soc. Series A*, **123**, 714–733.

Dirac, P. A. M. (1930) *Proc. Camb Phil Soc.*, **26**, 376.

Dunning, T. H. Jr. (1975) *J. Chem. Phys.*, **55**, 716–723.

Dykstra, C. E, Augspurger, J. D., Kirtman, B. and Malik, D. J. (1990) in *Reviews in Computational Chemistry*, (eds K. B. Lipkowitz and D. B. Boyd), VCH Publishers Ltd, New York.

Dykstra, C. E. and Liu, S-Y (1989) *Adv. Chem. Phys.*, **75**, 37.

Eyring, H., Walter, J. and Kimball, G. E. (1944) *Quantum Chemistry*, John Wiley & Sons Inc, New York.

Fermi, E. (1930) *Z. Phys.*, **60**, 320.

Finkelstein, B. and Horowitz, G. (1928) *Z. Phys.*, **48** 118.

Fischer, C. F. (1977) *The Hartree Fock Method for Atoms; A Numerical Approach*, John Wiley & Sons Inc, New York.

Frost, A. A, Rouse, R. A. and Vescelius, L. (1968) *Int. J. Quantum Chemistry IIS*, **43**.

Frost, A. A. (1967) *J. Chem. Phys.*, **47**, 3707; 3714.

Frost, A. A. (1968) *J. Phys. Chem.*, **72**, 1289.

Gelbart, W. M. (1974) *Adv. Chem. Phys.*, **XXVI**, 1.

Green, S. (1974) *Advances in Chemical Physics*, **25**, 179.

Hartree, D. R. (1927) *Proc. Camb Phil Soc.*, **24**, 89.

Hartree, D. R. (1957) *The Calculation of Atomic Structures*, John Wiley & Sons Inc, New York.

Hasenein, A. A. and Hinchliffe, A. (1990) in *Studies in Physical and Theoretical Chemistry*, Vol. 70, *Self Consistent Field Theory and Applications*, Elsevier, p. 670.

Hehre, W. J., Ditchfield, R., Radom, L. and Pople, J. A. (1970) *J. Amer. Chem. Soc.*, **92**, 4796.

Hehre, W. J., Stewart, R. F. and Pople, J. A. (1969) *J. Chem. Phys.*, **51**, 2657–2665.

Heitler, W. and London, F. (1927) *Z. Physik*, **44**, 455.

Henderson, J. B. (1967) *J. Amer. Chem. Soc.*, **89**, 7043.

Herman, F. and Skillman, S. (1963) *Atomic Structure Calculations*, Prentice Hall, Englewood Cliffs, NJ.

Herzberg, G. and Huber, K. P. (1979) Molecular Spectra and Molecular Structure 4, in *Constants of Diatomic Molecules*, van Nostrand, Princeton, NJ.

Hoffmann, R. (1963) *J. Chem. Phys.*, **39**, 1397–1412.

Hohenberg, P. and Kohn, W. (1964) *Phys. Rev.*, **B136**, 864–871.

Hückel, E. P. (1931) *Z. Physik*, **70**, 204; 310.

Humberto, J., Soscún, M., O'Malley, P. J. and Hinchliffe, A. *J. Mol. Struc. (THEOCHEM)*, paper in press 1995.

Kohn, W. and Sham, L. J. (1965) *Phys. Rev.*, **A140**, 1133–1138.

Koopmans, T. (1934) **Physica**, **1**, 104.

Latimer, W. M. and Rodebush, W. H. (1920) *J. Amer. Chem. Soc.*, **42**, 1419.

LeFèvre, R. J. W. (1938) *Dipole Moments*, Methuen.

Lipscomb, W. N. (1972) in *Trans Amer. Chem. Society*, **8**, 79.

Møller, Chr and Plesset, M. S. (1934) *Phy. Rev.*, **46**, 618.

March, N. H. (1992) *Electron Density Theory of Atoms and Molecules*, Academic Press, London.

McConnel, H. M. (1956) *J. Chem. Phys.*, **24**, 764.

McIntyre, E. F. and Hameka, H. F. (1978) *J. Chem. Phys.*, **68**, 3481.

McWeeny, R. and Sutcliffe, B. T. (1969) *Methods of Molecular Quantum Mechanics*, Academic Press, London.

Menédez, M. I., González, J. Sordo, J. A. and Sordo, T. L. (1994) *J. Mol. Struc. (THEOCHEM)*, **314**.

Morse, P. (1929) *Phys. Rev.*, **34**, 57.

Mulliken, R. S. (1955) *J. Chem. Phys.*, **23**, 1833–1840.
Mulliken, R. S., Rieke, C. A., Orloff, D. and Orloff, H. (1949) *J. Chem. Phys.*, **17**, 1248–1267.

Nishimoto, K. and Forster, L. S. (1966) *Theor Chim Acta*, **4**, 155–165.

O'Brien, E. F, Gutschick, V. P., McKoy, V. and McTague, J. P. (1973) *Phys. Rev.*, **A8**, 690.
Orville-Thomas, Quart, W. J (1957) *Reviews Chem. Soc.*, **11**, 162.

Pariser, R. and Parr, R. G. (1953) *J. Chem. Phys.*, **21**, 466–471.
Parr, R. G. and Yang, W. (1989) *Density-Functional theory of Atoms and Molecules*, Oxford University Press, Oxford.
Pearlman, R. S. (1987) *Chem. Design Automation News*, **2**, 1.
Poirer, R. Kari, R. and Csizmadia, I. G. (1985) *Handbook of Caussian Basic Sets*, Elsevier, Amsterdam.
Pople, J. A. (1953) *Transactions of the Faraday Society*, **49**, 1375–1385.
Pople, J. A. and Segal, G. A. (1965) *J. Chem. Phys.*, **43**, S129.
Pople, J. A. and Segal, G. A. (1966) *J. Chem. Phys.*, **44**, 3289.
Pople, J. A., Binkley, J. S. and Seeger, R. (1976) *Int. J. Quan. Chem.*, Symp No 10, 1.

Quantum Chemistry Literature Data Base (QCLDB), Supplement 13 (1993) Elsevier.

Renner, R. (1934) *Z. Physik*, **92**, 172.
Rogers, D. W. et al (1980) *J. Phys. Chem.*, **84**, 1810.
Roothaan, C. C. J. (1951) *J. Chem. Phys.*, **19**, 1445.
Roothaan, C. C. J. (1951a) *Revs Mod Phys.*, **23**, 69–89.

Rosen, N. (1931) *Phys. Rev.*, **38** 2099.
Rosenblum, B., Nethercot, Jr, A. H. and Townes, C. H. (1958) *Phys. Rev.*, **109**, 400.
Ruedenberg, K., Roothaan, C. C. J. and Jaunzemis, W. (1956) *J. Chem. Phys.*, **24**, 201.

Saunders, M., Houk, K. N., Wu, Yun-Dong, Still, Clark, W., Lipton, M., Chang, C., Guida, W. C. (1990) *J. Amer. Chem. Soc.*, **112**, 1416–1427.
Schlegel, H. (1989) *J. Comput. Chem.*, **10**, 209.
Schlegel, H. B. and Frisch, H. J. (1990) in *Theoretical and Computational Models for Organic Chemistry*, (eds F. J. Formoshina, I. G. Csizmadia and L. G. Arnaut), Kluiver, Dordrecht.
Schwarz, K. (1972) *Phys. Rev.*, **B5**, 2466.
Schwarz, K. and Connolly, J. W. D. (1971) *J. Chem. Phys.*, **55**, 4710.
Scrocco, E. and Tomasi, J. (19??) in *Topics in Current Chemistry*, 42, New Concepts II, Springer-Verlag, Berlin.
Slater (1965).
Slater J. C. (1929) *Phys. Rev.*, **34**, 1293.
Slater J. C. (1930) *Phys. Rev.*, **36**, 57.
Slater, J. C. (1951) *Phys. Rev.*, **81**, 385.
Slater, J. C. (1979) *The Calculation of Molecular Orbitals*, John Wiley & Sons, New York.
Smith, J. W. (1955) *Electric Dipole Moments*, Butterworths, London.
Snyder, L. C. and Basch, H. (1969) *J. Amer. Chem. Soc.*, **91**, 2189.
Stewart, J. J. P. (1989) *J. Comput. Chem.*, **10**, 209.
Streitwieser, A. (1961) *Molecular Orbital Theory for Organic Chemists*, John Wiley & Sons Inc, New York.
Su, W. P. and Schrieffer, J. R. (1980) *Proc. Natl. Acad. Sci. USA*, **77**, 5626.
Sugiura, Y. (1927) *Z. Physik*, **45**, 455.

Townes, C. H. and Dailey, B. P. (1949) *J. Chem. Phys.*, **17**, 782.
Trefler, M. and Gush, H. P. (1968) *Phys. Rev. Letters* **20**, 703.

Van Vleck, J. H. and Sherman, A. (1935) *Rev. Mod Phys.*, **7**, 167.

Weinbaum, S. (1933) *J. Chem. Phys.*, **1**, 317.
Weiner, P. K. and Kollman, P. A. (1981) *J. Comput. Chem.*, **2**, 287.
Weiss, S. and Leroi, G. (1968) *J. Chem. Phys.*, **48** 962.
Westheimer, F. H. and Meyer, J. E. (1946) *J. Chem. Phys.*, **14**, 733.
Wheland, G. W. and Pauling, L. (1959) *J. Amer. Chem. Soc.*, **57**, 3223.
Wilson, E. B. Jr., Decius, J. C. and Cross, P. C. (1955) *Molecular Vibrations. The Theory of Infrared and Raman Vibrational Spectra*, McGraw-Hill Book Company, NY.
Wind, H. (1965) *J. Chem. Phys.*, **42**, 2371.
Wolfsberg, M. and Helmhotte, L. (1952) *J. Chem. Phys.*, **20**, 837–843.

Index

Activity, pharmacological 22
AMBER forcefield 20, 208
Ampere 5
Amino acids 23
Angular momentum 152
Antisymmetry 44, 235
Antisymmetry factors 204, 206
Atomic Orbital, Slater (STO) 81
Atomic spectral data 88
Aufbau principle 2

Barrier, inversion 230
Basis function 61
   contracted 107
   diffuse 110
   gaussian 102ff
Basis set
   double ζ 100
   extended 108
   minimal 99
BECKE88 model 161
Bond separation 243
Bonding parameter 89
Born interpretation 1, 8, 48, 49
Born Oppenheimer approximation 26, 27,
   37, 61, 113, 167, 195, 199, 200
Brillouin theorem 83, 199, 202
Brookhaven protein data bank 20, 113
Brönsted acid site 245
Buckingham potential 223

Cambridge Structural Database 20
Charge density 123
CHARMm 20
Clausius-Mossotti 223

CNDO 87, 91, 92
Configuration Interaction (CI) 44ff
   CID 150
   CIS 145
   CISD 150
   monoexcited 143
CONCORD 20
Contour diagram 34
Correlation, electron 134, 233
Cotton-Mouton effect 202
Coulomb attraction 37
Counterpoise method 234

Debye, Peter 223
Density, electron spin 55
Density function 52, 54, 134
Density matrix 50ff
Density of States 156
Derivatives, energy 218
Determinant, Slater 65, 83
Dewar, M. J. S. 92ff
DFT 55
Dielectric Susceptibility 73
Diels-Alder reaction 190ff
Dipole derivatives 201
Dipole moment
   electric 49, 193, 197, 211, 212
   magnetic 197
Dirac P. A. M. 157
   –Slater exchange 157
Dissociation energy 169
   products 46
Docking
   unconstrained 23
   vector 23

Drude model   155
DTMM   21, 23, 112, 207ff

Eigenvalue equation   2, 70
Electron correlation   199
Electronic configuration   39
Electrostatic potential   207
Energy
    correlation   125, 134
    dissociation   31, 46
    interaction   143
    orbital   63
    variational   58, 104
    zero-point   188, 233
Energy minimization   170ff
Energy search
    conformational   176
    grid   170, 175
Equations
    secular   2
Equation of State
    ideal   222
    virial   222
Excited state
    double   135
    single   135

Fermi energy   156
Fermi contact term   237
Fermi correlation   134
Fermi level   156
Field
    electric   3, 225
    magnetic   225
    static   210
Field gradient
    electric   204ff
Finite differences   171
Finite Field SCF   218
Force
    Intermolecular   197
    Lorentz   6
Force constant   12, 17, 185
Force Field
    Urey-Bradley   18
    Valence   18, 180
Four vector   226
Function, Gaussian (GTO)   95

Gauge, choice of   226ff
GAUSSIAN   112, 160ff
Gaussian Type Orbital (GTO)   102ff
    energy optimized   108
    primitive   104, 106
Gauss' theorem   215

Geometry   122
    equilibrium   17
    optimization   229
Gradient   17
    vector   178

Hamiltonian   2
    CI   144
    dimensionless   30
    spin   239
Harmonic approximation   14, 168
Hartree Fock (SCF) model   33, 56, 67, 80
    limit   59, 124, 152
Hartree model   57, 154
Heitler London (VB) model   42
Hellmann-Feynman theorem   173
Heteroatoms   72
Hooke's Law   12ff
Hyperfine coupling constant   239

Indistinguishability   57
Inductive effect   78
Inner Shell   75ff
Integral
    canonical order   15
    electron repulsion   101
    one-electron   192
    overlap   30, 90
    transformation   137
    two-electron   43, 192
Interactions   193
    dispersion   125
    instantaneous   125
Isodesmic reaction   244
Isolated Binary Collision Method   224

Kerr effect   201, 214
Koopmans' theorem   64ff

Laplace's equation   211
LCAO   29, 68
Lennard Jones   19
Level of theory   112

Magnetizability   227
Mataga and Nishimoto   82
Matrix
    commutation   63
    coulomb   160
    exchange   160
    hessian   178, 190
    HF   62, 68
    overlap   68
    trace   62

MBER 201, 232
McConnell relation 241
Mechanics
    classical 11
    quantum 11
    relativistic 11
    wave 9
Minimization
    steepest descents 63
    global 173
Minimum
    global 173
    local 173, 182
MM2 20
MM3 20
MM(π) 71
MOBY1.5 20, 112, 208
Model
    Ab Initio 111
    Empirical 111
    Semi-Empirical 111
Møller-Plesset (Mpn) 13ff, 150
Momentum, generalized 226
Monte Carlo method 176
Muffin Tin potential 159
Mulliken, R.S. 74
Multipole Expansion 196ff

NDDO 92
Newton's Laws 12
Newton's Method 170ff, 179
Normalization 48

Occupation number 51, 158
Operator
    one-electron 37
    two-electron 37
Optimization
    Berny 181
    Fletcher-Powell 181
Orbital
    gaussian (GTO) 101
    ghost 234
    hybrid 86
    real equivalent 97
    Slater (STO) 85ff, 99, 101, 104
Orbital exponent 32, 98
Overlap matrix 51

Parameters
    atom 69
    bond 69
Particle
    distinguishable 40
    indistinguishable 41
Pauli's electron gas model 155

Penetration 88ff
Permittivity, relative 221
Perturbation expansion 138
Perturbation theory 137, 217ff
    self consistent 217
Point charges 3, 193
Polarizability
    anisotropy 214, 221
    atomic 215
    dipole 212ff
    incremental 224
    mean value 214
    principal values 213
Polarization
    dielectric 223
    induced 212
    spin 240
Polarization functions 100, 109
POLYATOM 109
Population analysis 52, 133
Population
    gross atomic 54
    net atomic 53
    overlap 53
Potential
    electric dipole 196
    electrostatic 4, 49
    Morse 15, 28
    vector 225
Potential energy 1, 13, 43, 58
    electrostatic 3
Potential energy curve 28
Potential energy surface 25, 38
Principal axes 195, 213
Principal values, 195, 213
Properties
    derivative 193
    frequency dependent 215
    induced 193, 210
    primary 192
Protein docking 22
Pulsed Nozzle FTIR 232

Quadrupole coupling constants 204ff
Quadrupole moment
    electric 194
    nuclear 204

Relativistic effects 199
Renner effect 27
Repulsion, electron 78
Roothaan, C.C.J. 63
Roothaan method 60
Rotation, restricted 228
Rules, Slater 98

Saddle Point 177ff
Schrödinger equation 2, 26, 28, 36ff, 41, 56, 155, 161
SECI 143
Second moments 194, 203
Self Consistent Field (SCF) 33, 56, 67, 80
Separation of Variables 38
Shielding (orbital) 98
Size consistency 143
Slater-Condon-Shortley Rules 57, 84
Slater determinant 47, 235
Slater type orbital (STO) 81
Soliton dynamics 73
Spectra
    depolarized 223
    microwave 229
    molecular electronic 91
Spheres 160
Spin 42
    electron 37, 41ff, 67, 128
    nuclear 203
Spin contamination 242
Spin density 128
Stationary point 186
Stark effect 197, 200
Steepest descents 179
Stern-Gerlach experiment 238
Structure
    covalent 47
    ionic 47
Supermolecule model 232
Surface plot 34ff
Symmetry 85, 122

Tamm-Dancoff approximation 143
Taylor Series 171, 211

Tensor properties 212
Thomas-Fermi model 157
Three body problem 56
Townes and Dailey model 205
Transition State 158, 178, 190

UHF model 66
Units
    electrostatic 6
    gaussian 6
    Hartree 7
    SI 5

Valence Bond (VB) 43
Valence State Ionization Energy 82
Variation Principle 2, 29, 45
Vibrations
    classical frequency 13
    coupled 8
Virial equation, dielectric 223
Visualization 33, 112

Wavefunction
    antisymmetric 42
    state 67
Wolfsberg-Helmholtz model 73
Work 4

Z-matrix 113ff
ZDO 80ff, 144
Zeeman effect 202, 238
Zeolite 245
Zero-order problem 138
Zero-point energy 169